Studies in Computational Intelligence

Volume 1063

D1824451

Series Editor

Janusz Kacprzyk, Polish Academy of Sciences, Warsaw, Poland

The series "Studies in Computational Intelligence" (SCI) publishes new developments and advances in the various areas of computational intelligence—quickly and with a high quality. The intent is to cover the theory, applications, and design methods of computational intelligence, as embedded in the fields of engineering, computer science, physics and life sciences, as well as the methodologies behind them. The series contains monographs, lecture notes and edited volumes in computational intelligence spanning the areas of neural networks, connectionist systems, genetic algorithms, evolutionary computation, artificial intelligence, cellular automata, self-organizing systems, soft computing, fuzzy systems, and hybrid intelligent systems. Of particular value to both the contributors and the readership are the short publication timeframe and the world-wide distribution, which enable both wide and rapid dissemination of research output.

This series also publishes Open Access books. A recent example is the book Swan, Nivel, Kant, Hedges, Atkinson, Steunebrink: The Road to General Intelligence https://link.springer.com/book/10.1007/978-3-031-08020-3

Indexed by SCOPUS, DBLP, WTI Frankfurt eG, zbMATH, SCImago.

All books published in the series are submitted for consideration in Web of Science.

Erik Cuevas · Omar Avalos · Jorge Gálvez

Analysis and Comparison of Metaheuristics

 Springer

Erik Cuevas
CUCEI
Universidad de Guadalajara
Guadalajara, Mexico

Omar Avalos
CUCEI
Universidad de Guadalajara
Guadalajara, Mexico

Jorge Gálvez
CUCEI
Universidad de Guadalajara
Guadalajara, Mexico

ISSN 1860-949X   ISSN 1860-9503 (electronic)
Studies in Computational Intelligence
ISBN 978-3-031-20107-3   ISBN 978-3-031-20105-9 (eBook)
https://doi.org/10.1007/978-3-031-20105-9

This Springer imprint is published by the registered company Springer Nature Switzerland AG
The registered company address is: Gewerbestrasse 11, 6330 Cham, Switzerland

Preface

From the optimization methods, there is a special set of approaches that are designed in terms of the interaction among the search agents of a group. Members inside the group cooperate to solve a global objective by using local accessible knowledge that is propagated through the set of members. With this mechanism, complex problems can be solved more efficiently than by considering the strategy of a single individual. In general terms, this group is referred to as a population, where individuals interact with each other in a direct or indirect manner by using local information from the environment. This cooperation among individuals produces an effective distributive strategy to solve problems. The most common term for such methods is metaheuristics. In general, there do not exist strict classifications of these methods. However, several kinds of algorithms have been coined depending on several criteria, such as the source of inspiration, cooperation among the agents, or type of operators.

The study of biological and social entities such as animals, humans, processes, or insects which manifest a cooperative behavior has produced several computational models in metaheuristic methods. Some examples of such agents include ants, particles, bees, teams, water drops, locust swarms, spiders, and bird flocks. In the group, each element conducts a simple strategy. However, due to its cooperative behavior, the final collective strategy produced by all individuals is normally very complex. The complex operation of all individuals is a consequence of the whole behavior among the agents generated during their interaction.

The complex operation of all elements cannot be reduced to the aggregation of behaviors of each agent in the group. The association of all simple agent behaviors is so complex that it usually is not easy to predict or deduce the global behavior of the whole population. Something remarkable is that these behavioral patterns appear without the existence of a coordinated control system but emerge from the exchange of local information among individuals. Therefore, there subsists a close relationship between individual and collective behavior. In general, the collective behavior of agents determines the behavior of the algorithm. On the other hand, collective behavior is also strongly influenced by the conditions under which each agent executes its operations.

The operations of each individual can modify its own behavior and the behavior of other neighbor agents, which also alters the global algorithm performance. Under such conditions, the most significant element of a metaheuristic scheme is the mechanism with which the interaction or cooperation among the agents is modeled. Cooperation among individuals that operate collectively happens in different mechanisms from which the exchange of information represents the most important. This process can be conducted through physical contact, visual information, audio messages, or chemical perceptual inputs. Examples of cooperation models presented in metaheuristic schemes are numerous; some examples include the dynamical task assignation performed in an ant colony without any central control or task coordination. The adoption of optimal spatial patterns builds by the self-organization in bird flocks and fish in schools. The hunting strategies were developed by predators. The purpose of computational metaheuristic schemes is to model the simple behaviors of agents and their local interactions with other neighboring agents to perform an effective search strategy for solving optimization problems. One example of these mechanisms is the Particle Swarm Optimization (PSO), which models two simple actions. Each agent (1) moves toward the best agent of the population, and (2) moves toward the position where the agent has reached its best location. As a consequence, the collective behavior of the group produces that all agents are attracted to the best positions experimented with by all elements. Another example is the ant colony optimization (ACO), which models the biological pheromone trail following the behavior of ants. Under this mechanism, each ant senses pheromone concentrations in its local position. Then, it probabilistically selects the path with the highest pheromone concentration. Considering this model, the collective effect in the population is to find the best option (shortest path) from a group of alternatives available in a decision-making problem.

There exist several features that clearly appear in most of the metaheuristic approaches, such as the use of diversification to force the exploration of regions of the search space, rarely visited until now, and the use of intensification or exploitation, to investigate some promising regions thoroughly. Another interesting feature is the use of memory to store the best solutions encountered. For these reasons, metaheuristics methods quickly became popular among researchers to solve simple to complex optimization problems in different areas.

Most of the problems in science, engineering, economics, and life can be translated as optimization or a search problem. According to their characteristics, some problems can be simple, so that they can be solved by traditional optimization methods based on mathematical analysis. However, most of the problems of practical importance, such as system identification, parameter estimation, energy systems, etc., represent conflicting scenarios that are very hard to be solved by using traditional approaches. Under such circumstances, metaheuristic algorithms have emerged as the best alternative to solve this kind of complex formulation. Therefore, metaheuristic techniques have consolidated as a very active research subject in the last ten years. During this time, various new metaheuristic approaches have been introduced. They have been experimentally examined on a set of artificial benchmark problems and in a large number of practical applications. Although metaheuristic methods represent one of the most exploited research paradigms in computational intelligence, there are still a large number of open challenges. They range from premature convergence, inability to maintain population diversity, and the combination of metaheuristic paradigms with other algorithmic schemes toward extending the available techniques to tackle ever more difficult problems.

Numerous books have been published taking into account any of the most widely known metaheuristic methods, namely ant colony algorithms and particle swarm optimization but attempts to consider the discussion of new alternative approaches are always scarce. Initial metaheuristic schemes maintain in their design several limitations such as premature convergence and the inability to maintain population diversity. Recent metaheuristic methods have addressed these difficulties providing, in general, better results. Many of these novel swarm approaches have also been introduced lately. In general, they propose new models and innovative cooperation models for producing an adequate exploration and exploitation of large search spaces considering a significant number of dimensions. Most of the new metaheuristics present promising results. Nevertheless, they are still in their initial stage. To grow and attain their complete potential, new metaheuristic methods must be applied in a great variety of problems and contexts so that they not only perform well in their reported sets of optimization problems, but also in new complex formulations. The only way to accomplish this is by making possible the transmission and presentation of these methods in different technical areas as optimization tools. In general, once a scientific engineer or practitioner recognizes a problem as a particular instance of a more generic class, he/she can select one of the different metaheuristic algorithms that guarantee an expected optimization performance. Unfortunately, the set of options is concentrated in algorithms whose popularity and high proliferation are better than the new developments.

The excessive publication of developments based on the simple modification of popular metaheuristic methods presents an important disadvantage: They avoid the opportunity to discover new potential techniques and procedures which can be useful to solve problems formulated by the academic and industrial communities. In the last years, several promising metaheuristic schemes that consider very interesting concepts and operators have been introduced. However, they seem to have been completely overlooked in the literature in favor of the idea of modifying, hybridizing, or restructuring popular metaheuristic approaches.

The goal of this book is to present a comparative perspective of current metaheuristic developments, which have proved to be effective in their application to several complex problems. The book considers different metaheuristic methods and their practical applications. This structure is important to us because we recognize this methodology as the best way to assist researchers, lecturers, engineers, and practitioners in the solution of their own optimization problems.

This book has been structured so that each chapter can be read independently from the others. Chapter 1 describes the main characteristics and properties of metaheuristic methods. This chapter analyses the most important concepts of metaheuristic schemes.

Chapter 2 presents the comparison of various metaheuristic techniques currently in use applied to the design of 2D-IIR digital filters. The design of two-dimensional Infinite Impulse Response (2D-IIR) filters has recently attracted attention in several areas of engineering because of their wide range of applications. Synthesizing a user-defined filter in a 2D-IIR structure can be interpreted as an optimization problem. However, since 2D-IIR filters can easily produce unstable transfer functions, they tend to produce multimodal error surfaces whose cost functions are significantly complex to optimize. On the other hand, metaheuristic algorithms are well-known global optimization methods with the capacity to explore complex search spaces for a suitable solution. Each metaheuristic technique holds distinctive characteristics to appropriately fulfill the requirements of particular problems. Therefore, no single metaheuristic algorithm can solve all problems competitively. In order to know the advantages and limitations of metaheuristic methods, their correct evaluation is an important task in the computational intelligence community. Furthermore, metaheuristic algorithms are stochastic processes with random operations. Under such conditions, for obtaining significant conclusions, appropriate statistical methods must be considered. Although several comparisons among metaheuristic methods have been reported in the literature, their conclusions are based on a set of synthetic functions without considering the application context or appropriate statistical treatment.

Chapter 3 exhibits a comparative study of the most used evolutionary techniques used to solve the parameter estimation problem of chaotic systems not only based on the performance criteria but also on their solution homogeneity using statistical analysis. Results over the Lorenz and Chen chaotic systems are analyzed and statistically validated using non-parametric tests. In recent years, Parameter Estimation (PE) has attracted the attention of the scientific community. It could be applied in different fields of engineering and science. An important area of research is the system identification of Chaotic Systems (CS) in order to synchronize and control chaos. The parameter estimation of CS is a highly nonlinear optimization problem within a multi-dimensional space where classic optimization techniques are not suitable to use. Metaheuristic Computation Techniques (MCT) are commonly used to solve complex nonlinear optimization problems. Recently, some classic and modern MCTs have been proposed in order to estimate the parameters for chaotic systems; nevertheless, the results and conclusions reported in the literature are based only on the cost function values of each MCT regardless of their solutions.

Chapter 4 presents a comparative analysis of the application of five recent metaheuristic schemes to the shape recognition problem such as the Grey Wolf Optimizer (GWO), Whale Optimizer Algorithm (WOA), Crow Search Algorithm (CSA), Gravitational Search Algorithm (GSA), and Cuckoo Search (CS). Since such approaches have been successful in several new applications, the objective is to determine their efficiency when they face a complex problem such as shape detection. Numerical simulations, performed on a set of experiments composed of images with different difficulty levels demonstrate the capacities of each approach. Shape recognition in images represents one of the complex and hard-solving problems in computer vision due to its nonlinear, stochastic, and incomplete nature. Classical image processing techniques have been normally used to solve this problem. Alternatively, shape recognition has also been conducted through metaheuristic algorithms. They have demonstrated to have a competitive performance in terms of robustness and accuracy. However, all of these schemes use old metaheuristic algorithms as the basis to identify geometrical structures in images. Original metaheuristic approaches experiment with several limitations, such as premature convergence and low diversity. Through the introduction of new models and evolutionary operators, recent metaheuristic methods have addressed these difficulties providing in general better results.

Chapter 5 presents the comparison of various evolutionary computation optimization techniques applied to IIR model identification. In the comparison, special attention is paid to recently developed algorithms such as Cuckoo Search and Flower Pollination Algorithm, including also popular approaches. Results over several models are presented and statistically validated.

In Chap. 6, it is presented an algorithm for parameter identification of fractional-order chaotic systems. In order to determine the parameters, the proposed method uses a novel evolutionary method called Locust Search (LS), which is based on the behavior of swarms of locusts. Different from most existent evolutionary algorithms, it explicitly avoids the concentration of individuals in the best positions, avoiding critical flaws such as premature convergence to suboptimal solutions and the limited exploration-exploitation balance. Numerical simulations have been conducted on the fractional-Order Van der Pol oscillator to show the effectiveness of the proposed scheme.

In Chap. 7, a comparative study between metaheuristic techniques used for solar cells parameter estimation is proposed, using the one-diode, two-diode, and three-diode models. This study presents three solar cell models with different operating conditions; the results obtained are presented and statistically validated. The use of renewable energy has increased in recent years due to the environmental consequences of fossil fuel employment. Many alternatives have been proposed for the exploitation of clean energy; one of the most used is solar cells due to their unlimited source of power. Solar cell parameter estimation has become a critical task for many research areas nowadays. The efficiency of solar cell operation depends on the parameters in their design, which is a complex task for the nonlinearity and the multimodal error surface generated, hindering their minimization. Metaheuristic computation techniques (MCT) are employed to determine competitive solutions to complex optimization problems. However, when a new MCT is developed, it is tested using well-known functions with an exact solution without considering the real-world application, which normally is nonlinear and with unknown behavior. Some MCTs have been used to determine the parameters of the solar cell. However, there is an important limitation that all of them share. Frequently, they obtain suboptimal solutions due to the inappropriate balance between exploration and exploitation of their search strategy.

Finally, Chap. 8 conducts an experimental study where it is analyzed the performance between agent-based models and metaheuristic approaches. Agent-based models represent new approaches to characterize systems through simple rules. Under such techniques, complex global behavioral patterns emerge from the agent interactions produced by the rules. Recently, due to their capacities, metaheuristic methods have attracted the attention of the optimization community. Even though these approaches are built to emulate very distinct phenomena, their structure and operators are very similar. Despite their diversity, several metaheuristic methods recycle the same elements from other metaheuristic techniques that have been demonstrated to be efficient. These elements have been designed without considering the produced final search patterns. Contrarily, agent-based modeling aims to relate the global behavioral patterns produced by the collective interaction of the individuals with the set of rules that describe their behavior. In this chapter, we remark on the association between metaheuristic elements and agent-based models. Therefore, different agent-based structures that produce interesting search patterns can be employed to generate promising optimization methods. To demonstrate the abilities of this methodology, an agent-based approach known as "Heroes and Cowards" has been structured as a metaheuristic technique. This agent-based model implements a small set of rules to generate two search patterns that can perform as exploration and exploitation stages. To evaluate its performance, this algorithm has been compared with several metaheuristic methods.

As authors, we wish to thank many people who were somehow involved in the writing process of this book. We express our gratitude to Prof. Janusz Kacprzyk, who so warmly sustained this project.

Guadalajara, Mexico   Erik Cuevas
                                                                                           Omar Avalos
                                                                                           Jorge Gálvez

Contents

Chapter 1
Fundamentals of Metaheuristic Computation

This chapter presents the main concepts of metaheuristic schemes. The objective of this chapter is to introduce the characteristics and properties of these approaches. An important purpose of this chapter is also to recognize the importance of metaheuristic methods to solve optimization problems in cases in which traditional techniques are not suitable.

1.1 Formulation of an Optimization Problem

Most industrial and engineering systems require the use of an optimization process for their operation. In such systems, it is necessary to find a specific solution that is considered the best in terms of a cost function. In general terms, an optimization scheme corresponds to a search strategy that has an objective to obtain the best solution considering a set of potential alternatives. This bets solution represents the best possible solution that, according to the cost function, solves the optimization formulation appropriately [1].

Consider a public transportation system of a specific town, for illustration proposes. In this example, it is necessary to find the "best" path to a particular target destination. To assess each possible alternative and then get the best possible solution, an adequate criterion should be taken into account. A practical criterion could be the relative distances among all possible routes. Therefore, a hypothetical optimization scheme selects the option with the smallest distance as a final output. It is important to recognize that several evaluation elements are also possible, which could consider other important criteria such as the number of transfers, the time required to travel from one location to another, or ticket price.

Optimization can be formulated as follows: Consider a function $f : S \to \Re$ which is called the cost function, find the argument that minimizes f:

$$x^* = \arg \min_{x \in S} f(\mathbf{x}) \tag{1.1}$$

© The Author(s), under exclusive license to Springer Nature Switzerland AG 2023
E. Cuevas et al., *Analysis and Comparison of Metaheuristics*, Studies in Computational Intelligence 1063, https://doi.org/10.1007/978-3-031-20105-9_1

S corresponds to the search space that refers to all possible solutions. In general terms, each possible solution solves in a different quality the optimization problem. Commonly, the unknown elements of x represent the decision variables of the optimization formulation. The cost function f determines the quality of each candidate solution. It evaluates the way in which a candidate element x solves the optimization formulation.

In the example of public transportation, S represents all subway stations, bus lines, etc., available in the database of the transportation system. $\mathbf{x}$ represents a possible path that links the start location with the final destination. $f(\mathbf{x})$ is the cost function that assesses the quality of each possible route. Some other constraints can be incorporated as a part of the problem definition, such as the ticket price or the distance to the destination (in different situations, it is taken into account the combination of both indexes, depending on our preferences).

When additional constraints exist, the optimization problem is called constrained optimization (different from unconstrained optimization where such restrictions are not considered). Under such conditions, an optimization formulation involves the next elements:

- One or several decision variables from $\mathbf{x}$, which integrate a candidate solution
- A cost function $f(\mathbf{x})$ that evaluates the quality of each solution $\mathbf{x}$
- A search space S that defines the set of all possible solutions to $f(\mathbf{x})$
- Constraints that represent several feasible regions of the search space S.

In practical terms, an optimization approach seeks within a search space S a solution for $f(\mathbf{x})$ in a reasonable period of time with enough accuracy. The performance of the optimization method also depends on the type of formulation. Therefore, an optimization problem is well-defined if the following conditions are established:

1. There is a solution or set of solutions that satisfy the optimal values.
2. There is a specific relationship between a solution and its position so that small displacements of the original values generate light deviations in the objective function $f(\mathbf{x})$.

1.2 Classical Optimization Methods

Once an engineering problem has been translated into a cost function, the next operation is to choose an adequate optimization method. Optimization schemes can be divided into two sets: classical approaches and metaheuristic methods [2].

Commonly, $f(\mathbf{x})$ presents a nonlinear association in terms of its modifiable decision variables $\mathbf{x}$. In classical optimization methods, an iterative algorithm is employed to analyze the search space efficiently. Among all approaches introduced in the literature, the methods that use derivative-descent principles are the most popular. Under such techniques, the new position x_{k+1} is determined from the current location x_k in a direction toward $\mathbf{d}$:

$$x_{k+1} = x_k + \alpha \mathbf{d}, \tag{1.2}$$

where α symbolizes the learning rate that determines the extent of the search step in the direction to $\mathbf{d}$. The direction $\mathbf{d}$ in Eq. 1.2 is computed, assuming the use of the gradient ($\mathbf{g}$) of the objective function $f(\cdot)$.

One of the most representative methods of the classical approaches is the steepest descent scheme. Due to simplicity, this technique allows us to solve efficiently objective functions. Several other derivative-based approaches consider this scheme as the basis for the construction of more sophisticated methods. The steepest descent scheme is defined under the following formulation:

$$x_{k+1} = x_k - \alpha \mathbf{g}(f(x)), \tag{1.3}$$

In spite of its simplicity, classical derivative-based optimization schemes can be employed as long as the cost function presents two important constraints:

- (I) The cost function can be two-timed derivable.
- (II) The cost function is unimodal; i.e., it presents only one optimal position.

The optimization problem defines a simple case of a derivable and unimodal objective function. This function fulfills the conditions (I) and (II):

$$f(x_1, x_2) = 10 - e^{-(x_1^2 + 3 \cdot x_2^2)} \tag{1.4}$$

Figure 1.1 shows the function defined by formulation 1.4.

Considering the current complexity in the design of systems, there are too few cases in which traditional methods can be applied. Most of the optimization problems imply situations that do not fulfill the constraints defined for the application of gradient-based methods. One example involves combinatorial problems where there is no definition of differentiation. There exist also many situations why an optimization problem could not be differentiable. One example is the "floor" function, which delivers the minimal integer number of its argument. This operation applied in Eq. 1.4 transforms the optimization problem from (Eq. 1.5) one that is differentiable to others not differentiable. This problem can be defined as follows (Fig. 1.2):

$$f(x_1, x_2) = \text{floor}\left(10 - e^{-(x_1^2 + 3 \cdot x_2^2)}\right) \tag{1.5}$$

Although an optimization problem can be differentiable, there exist other restrictions that can limit the use of classical optimization techniques. Such a restriction corresponds to the existence of only an optimal solution. This fact means that the cost function cannot present any other prominent local optima. Let us consider the minimization of the Griewank function as an example.

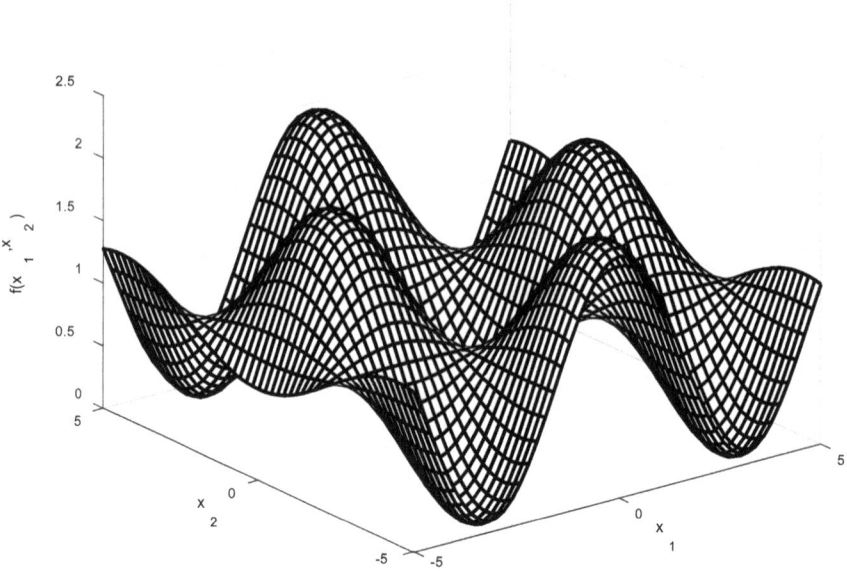

Fig. 1.1   Cost function with unimodal characteristics

$$\text{minimize } f(x_1, x_2) = \frac{x_1^2 + x_2^2}{4000} - \cos(x_1)\cos\left(\frac{x_2}{\sqrt{2}}\right) + 1$$
$$\text{subject to} \qquad \begin{array}{c} -5 \le x_1 \le 5 \\ -5 \le x_2 \le 5 \end{array} \qquad\qquad (1.6)$$

A close analysis of the formulation presented in Eq. 1.6; it is clear that the optimal global solution is located in $x_1 = x_2 = 0$. Figure 1.3 shows the cost function established in Eq. 1.6. As can be seen from Fig. 1.3, the cost function presents many local optimal solutions (multimodal) so that the gradient-based techniques with a randomly generated initial solution will prematurely converge to one of them with a high probability.

Considering the constraints of gradient-based approaches, it makes difficult their use to solve a great variety of optimization problems in engineering. Instead, some other alternatives which do not present restrictions are needed. Such techniques can be employed in a wide range of problems [3].

1.3   Metaheuristic Computation Schemes

Metaheuristic [4] schemes are derivative-free methods, which do not need that the cost function maintains the restrictions of being two-timing differentiable or

$$f(x_1, x_2) = \text{floor}\left(10 - e^{-\left(x_1^2 + 3 \cdot x_2^2\right)}\right)$$

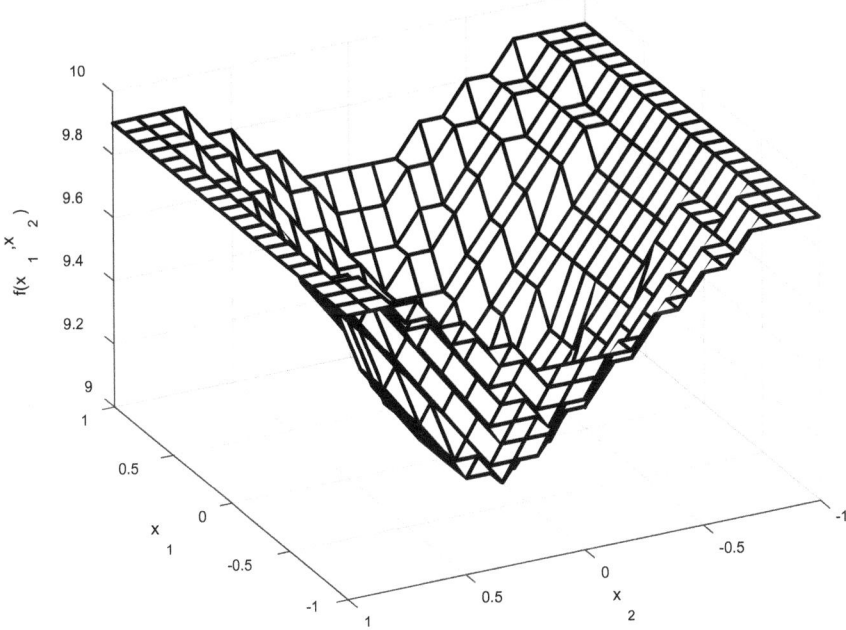

Fig. 1.2 A non-differentiable produced through the use of the floor function

unimodal. Under such conditions, metaheuristic methods represent global optimization methods which can deal with several types of optimization problems such as non-convex, nonlinear, and multimodal problems subject to linear or nonlinear constraints with continuous or discrete decision variables.

The area of metaheuristic computation maintains a rich history. With the demands of more complex industrial processes, it is necessary for the development of new optimization techniques that do not require prior knowledge (hypotheses) on the optimization problem. This lack of assumptions is the main difference between classical gradient-based methods. In fact, the majority of engineering system applications are highly nonlinear or characterized by noisy objective functions. Furthermore, in several cases, there is no explicit deterministic expression for the optimization problem. Under such conditions, the evaluation of each candidate solution is carried out through the result of an experimental or simulation process. In this context, the metaheuristic methods have been proposed as optimization alternatives.

A metaheuristic approach is a generic search strategy used to solve optimization problems. It employs a cost function in an abstract way, considering only its

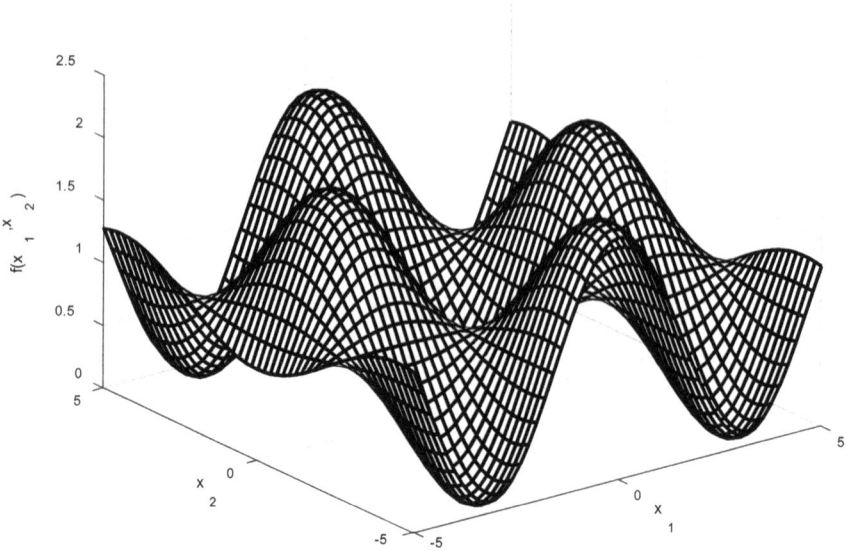

Fig. 1.3 The Griewank function with multimodal characteristics

evaluations in particular positions without considering its mathematical properties. Metaheuristic methods do not need any hypothesis on the optimization problem nor any kind of prior knowledge on the objective function. They consider the optimization formulation as "black boxes" [5]. This property is the most prominent and attractive characteristic of metaheuristic computation.

Metaheuristic approaches collect the necessary knowledge about the structure of an optimization problem by using the information provided by all solutions (i.e., candidate solutions) assessed during the optimization process. Then, this knowledge is employed to build new candidate solutions. It is expected that these new solutions present better quality than the previous ones.

Currently, different metaheuristic approaches have been introduced in the literature with good results. These methods consider modeling our scientific knowledge of biological, natural, or social systems, which, under some perspective, can be understood as optimization problems [6].

These schemes involve the cooperative behavior of bee colonies such as the Artificial Bee Colony (ABC) technique [8], the social behavior of bird flocking and fish schooling such as the Particle Swarm Optimization (PSO) algorithm [7], the emulation of the bat behavior such as the Bat Algorithm (BA) method [10], the improvisation process that occurs when a musician searches for a better state of harmony such as the Harmony Search (HS) [9], the social-spider behavior such as the Social Spider Optimization (SSO) [12], the mating behavior of firefly insects such as the Firefly (FF) method [11], the emulation of immunological systems as the clonal selection algorithm (CSA) [14], the simulation of the animal behavior in a group such as the Collective Animal Behavior [13], the emulation of the differential and conventional evolution in species such as the Differential Evolution (DE) [16], the simulation of the electromagnetism phenomenon as the electromagnetism-Like algorithm [15] and Genetic Algorithms (GA) [17], respectively.

1.4 Generic Structure of a Metaheuristic Method

In general terms, a metaheuristic scheme refers to a search strategy that emulates under a particular point of view a specific biological, natural or social system. A generic metaheuristic method involves the following characteristics:

1. Maintain a population of candidate solutions.
2. This population is dynamically modified through the production of new solutions.
3. A cost function associates the capacity of a solution to survive and reproduce similar elements.
4. Different operations are defined in order to explore an appropriately exploit the space of solutions through the production of new promising solutions.

Under the metaheuristic methodology, it is expected that, on average, candidate solutions enhance their quality during the evolution process (i.e., their ability to solve the optimization formulation). In the operation of the metaheuristic scheme, the operators defined in its structure will produce new solutions. The quality of such solutions will be improved as the number of iterations increases. Since the quality of each solution is associated with its capacity to solve the optimization problem, the metaheuristic method will guide the population towards the optimal global solution. This powerful mechanism has allowed the use of metaheuristic schemes to several complex engineering problems in different domains [18–22].

Most of the metaheuristic schemes have been devised to solve the problem of finding a global solution of a nonlinear optimization problem with box constraints in the following form:

$$\text{Maximize/Minimize } f(\mathbf{x}), \mathbf{x} = (x_1, \ldots, x_d) \in \Re^d$$
$$\text{subject to} \qquad \mathbf{x} \in \mathbf{X} \tag{1.7}$$

where $f : \Re^d \rightarrow \Re$ represents a nonlinear function whereas $\mathbf{X} = \{\mathbf{x} \in \Re^d | l_i \leq x_i \leq u_i, i = 1, \ldots, d\}$ corresponds to the feasible search space, restricted by the lower (l_i) and upper (u_i) bounds.

With the objective of solving the problem of Eq. 1.6, under the metaheuristic computation methodology, a group (population) $\mathbf{P}^k$ ($\{\mathbf{p}_1^k, \mathbf{p}_2^k, \ldots, \mathbf{p}_N^k\}$) of N possible solutions (individuals) is modified from a start point ($k = 0$) to a total gen number iterations ($k = gen$). In the beginning, the scheme starts initializing the set of N candidate solutions with random values uniformly distributed between the pre-specified lower (l_i) and upper (u_i) limits. At each generation, a group operations are used over the

Fig. 1.4 Generic procedure
of a metaheuristic scheme

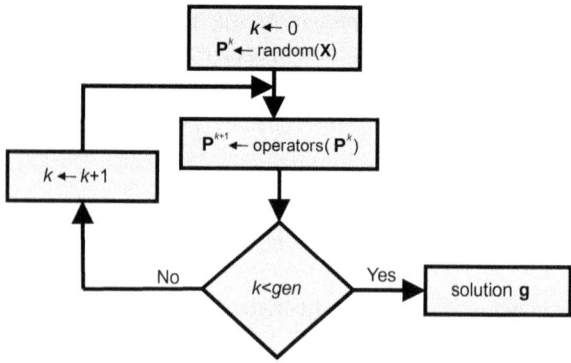

current population $\mathbf{P}^k$ to generate a new set of individuals $\mathbf{P}^{k+1}$. Each possible solution $\mathbf{p}_i^k$ ($i \in [1, \ldots, N]$) corresponds to a d-dimensional vector $\{p_{i,1}^k, p_{i,2}^k, \ldots, p_{i,d}^k\}$ where each element represents a decision variable of the optimization problem to be solved. The capacity of each possible solution $\mathbf{p}_i^k$ to solve the optimization problem is assessed by considering a cost function $f\left(\mathbf{p}_i^k\right)$ whose delivered value symbolizes the fitness value of $\mathbf{p}_i^k$. As the evolution process progresses, the best solution $\mathbf{g}$ $(g_1, g_2, \ldots g_d)$ seen so-far is maintained since it is the best available solution. Figure 1.4 shows an illustration of the generic procedure of a metaheuristic method.

References

1. Akay B, Karaboga D (2015) A survey on the applications of artificial bee colony in signal, image, and video processing. SIViP 9(4):967–990
2. Yang X-S (2010) Engineering optimization. Wiley, Inc.
3. Treiber MA (2013) Optimization for computer vision an introduction to core concepts and methods. Springer
4. Simon D (2013) Evolutionary optimization algorithms. Wiley
5. Blum C, Roli A (2003) Metaheuristics in combinatorial optimization: overview and conceptual comparison. ACM Comput Surv (CSUR) 35(3):268–308. https://doi.org/10.1145/937503.937505
6. Nanda SJ, Panda G (2014) A survey on nature inspired metaheuristic algorithms for partitional clustering. Swarm Evol Comput 16:1–18
7. Kennedy J, Eberhart R (1995) Particle swarm optimization. In: Proceedings of the 1995 IEEE international conference on neural networks, vol 4, December 1995, pp 1942–1948
8. Karaboga D (2005) An idea based on honey bee swarm for numerical optimization. Technical report-TR06. Engineering Faculty, Computer Engineering Department, Erciyes University
9. Geem ZW, Kim JH, Loganathan GV (2001) A new heuristic optimization algorithm: harmony search. Simulations 76:60–68
10. Yang XS (2010) A new metaheuristic bat-inspired algorithm. In: Cruz C, González J, Krasnogor GTN, Pelta DA (eds) Nature inspired cooperative strategies for optimization (NISCO 2010), studies in computational intelligence, vol 284. Springer, Berlin, pp 65–74
11. Yang XS (2009) Firefly algorithms for multimodal optimization. In: Stochastic algorithms: foundations and applications, SAGA 2009, Lecture notes in computer sciences, vol 5792, pp 169–178

12. Cuevas E, Cienfuegos M, Zaldívar D, Pérez-Cisneros M (2013) A swarm optimization algorithm inspired in the behavior of the social-spider. Expert Syst Appl 40(16):6374–6384
13. Cuevas E, González M, Zaldivar D, Pérez-Cisneros M, García G (2012) An algorithm for global optimization inspired by collective animal behavior. Discrete Dyn Nat Soc, art. no. 638275
14. de Castro LN, von Zuben FJ (2002) Learning and optimization using the clonal selection principle. IEEE Trans Evol Comput 6(3):239–251
15. Birbil ŞI, Fang SC (2003) An electromagnetism-like mechanism for global optimization. J Glob Optim 25(1):263–282
16. Storn R, Price K (1995) Differential evolution—a simple and efficient adaptive scheme for global optimisation over continuous spaces. Technical report TR-95–012, ICSI, Berkeley, CA
17. Goldberg DE (1989) Genetic algorithm in search optimization and machine learning. Addison-Wesley
18. Cuevas E (2013) Block-matching algorithm based on harmony search optimization for motion estimation. Appl Intell 39(1):165–183
19. Díaz-Cortés M-A, Ortega-Sánchez N, Hinojosa S, Cuevas E, Rojas R, Demin A (2018) A multi-level thresholding method for breast thermograms analysis using Dragonfly algorithm. Infrared Phys Technol 93:346–361
20. Díaz P, Pérez-Cisneros M, Cuevas E, Hinojosa S, Zaldivar D (2018) An improved crow search algorithm applied to energy problems. Energies 11(3):571
21. Cuevas E, González A, Fausto F, Zaldívar D, Pérez-Cisneros M (2015) Multithreshold segmentation by using an algorithm based on the behavior of Locust Swarms. Math Probl Eng 2015:805357
22. Hinojosa S, Oliva D, Cuevas E … Avalos O, Gálvez J (2018) Improving multi-criterion optimization with chaos: a novel multi-objective chaotic crow search algorithm. Neural Comput Appl 29(8):319–335

Chapter 2
A Comparative Approach for Two-Dimensional Digital IIR Filter Design Applying Different Evolutionary Computational Techniques

The two-dimensional Infinite Impulse Response (2D-IIR) filters design has attracted attention in many fields of engineering due to its wide range of applications. Incorporating a user-defined filter in the 2D-IIR structure can be represented as an optimization problem. Nevertheless, considering that 2D-IIR filters can easily generate unstable transfer functions, they produce multimodal error surfaces which are complex to optimize. On the other hand, Evolutionary Computation (EC) techniques methods with the ability to explore complex search spaces for suitable solutions. Each EC method explores distinctive features to properly fulfill the requirements of special problems. In order to identify the advantages and weaknesses of EC techniques, their correct evaluation is an essential task in the computational intelligence community. Moreover, EC algorithms consider stochastic processes with random operations. Under such circumstances, to obtain significant conclusions, an appropriate statistical analysis must be considered. Although many comparisons between EC techniques are reported in the literature, their results are based on sets of synthetic functions, without analyzing the application or an appropriate statistical procedure. This chapter introduces a comparative study of various EC methods applied to 2D-IIR digital filters design. The experimental results are presented, and statistical analysis is carried out.

2.1 Introduction

2D digital filters have been used in several engineering fields, such as computer vision and signal processing, just to mention a few [1–3]. 2D filters can be classified into two principal groups considering their impulse response: Infinite impulse response (IIR) and finite impulse response (FIR). Generally, the IIR filters produce a better performance than the FIR models considering even fewer parameters in their structure. Nevertheless, since 2D IIR filters can produce unstable responses, it is difficult to design them compared with their analogs [4].

© The Author(s), under exclusive license to Springer Nature Switzerland AG 2023
E. Cuevas et al., *Analysis and Comparison of Metaheuristics*, Studies in Computational Intelligence 1063, https://doi.org/10.1007/978-3-031-20105-9_2

The main purpose of a filter design is the correct approximation to the response for a determined filter usually provided by the user requirements. The methodologies for 2D-IIR filter design can be classified into two categories [5, 6]; transformation techniques and optimization-based methods. In transformation techniques, the 2D filter design is generated from a one-dimensional prototype that satisfies the desired response. Under such a scheme, the filter design is faster but adversely with low accuracy. On the other hand, in optimization-based methods, the 2D-IIR filter design is represented as an optimization task. Hence, evolutionary techniques are employed to determine the best filter design over a set of solutions. Due to their synthesizing problems, the 2D-IIR filter design is normally carried through evolutionary computational techniques instead of analytical approaches [7]. Under the optimization methodology, the filter design is highly dependent on the employed optimization technique. Usually, the generation of a filter through an optimization procedure is computationally high and takes more time than transformation methods.

In optimization techniques, the design process consists of determining the best parameters of the 2D-IIR filter that fit better with the desired response. To perform the optimization process, an objective function J is used to determine the similarity between the designed filter and the desired response. Nevertheless, considering that 2D-IIR filters can generate unstable transfer functions, they normally produce multimodal error surfaces that are significantly complex to optimize [8]. These kinds of objective functions make challenging the procedure of derivative-based methods since they can be easily trapped into a suboptimal solution [9]. Under such circumstances, techniques that use the gradient descent method as a basic algorithm are hardly considered for the 2D-IIR filter design [10].

On the other hand, evolutionary computational techniques (ECT) are optimization approaches motivated by our perception of biological or social systems, which can be considered search strategies at some abstraction level [11]. Some examples of common ECT techniques are the Particle Swarm Optimization (PSO) [12], Differential Evolution (DE) [13], Artificial Bee Colony (ABC) algorithm [14], Harmony Search (HS) [15], Flower Pollination Algorithm (FPA) [16], and Gravitational Search Algorithm (GSA) [17].

The performance of the ECT approaches to determine a good solution depends on its capability of balancing the exploitation and the exploration of the search space [18]. The exploration process consists of the inspection of new positions while the exploitation describes the process of bounded search in previously visited locations for improving their quality. Performing mainly exploration deteriorates the accuracy of the search process but enhances its capacity to determine new potential solutions [19]. On the other hand, if the exploitation process is encouraged, then, previous solutions are largely improved obtaining sub-optimal solutions [20]. Moreover, each optimization problem commonly requires a different trade-off between these concepts [21]. Since the exploration–exploitation balance is a big issue in the community, each ECT approach determines an experimental solution by combining deterministic rules and stochastic operators. For this reason, there is not an ECT that solves all problems competitively.

To know the strengths and weaknesses of EC techniques, their correct evaluation is an essential task in the computational intelligence society. Under such circumstances, suitable statistical techniques must be included for determining notable conclusions. Although many comparative studies among EC techniques have been described in the literature [22–24] where their conclusions are based on a set of well-known test functions, without considering an appropriate statistical procedure. This chapter presents the comparison of different EC approaches by solving the complex application of 2D-IIR filter design. In the study, particular consideration is paid to the most popular EC techniques currently in use such as PSO, ABC, DE, HS, and GSA. In the experimentation stage, the statistical validation of Wilcoxon and Bonferroni correction have been carried out.

2.2 Evolutionary Computation Algorithms

Evolutionary computation (ECT) algorithms are robust mechanisms for solving difficult optimization tasks. Distinct from classical schemes, ECT techniques are mainly developed to deal with multimodal objective functions. Most of the ECT algorithms are motivated by biological or social systems, which can be considered search strategies at some abstraction level. Generally, ECT methods are conformed by two main elements; determinist operations and stochastic sequences. Combining both elements, each ECT algorithm determines a specific solution to the exploration–exploitation balance. Under such circumstances, each EC technique contains unique features that appropriately fulfill the specifications of specific problems.

Different ECT methods have been introduced in the literature. According to their applications [11], the algorithms Particle Swarm Optimization (PSO) [12], Differential Evolution (DE) [13], Artificial Bee Colony (ABC) algorithm [14], Harmony Search (HS) [15], Flower Pollination Algorithm (FPA) [16], and Gravitational Search Algorithm (GSA) [17] are considered the most common currently in use. Therefore, they are employed in our comparative study. In this subsection, a brief description of each method used in the comparison is presented.

2.2.1 Particle Swarm Optimization (PSO)

Probably the PSO is the most popular optimization method. It was introduced in 1995 by Kennedy [12] and it was inspired by the behavior of bird flocking. This technique employs a group of particles of N elements $\mathbf{p}^k$ ($\{\mathbf{p}_1^k, \mathbf{p}_2^k, ..., \mathbf{p}_N^k\}$) as potential solutions, which are evolved from an initial position ($k = 0$) until reaching a certain stop criterion (Normally a total number of iterations). ($k = gen$) Each element $\mathbf{p}_i^k$ ($i \in [1, ..., N]$) represents a d-dimensional vector $\{\mathbf{p}_{i,1}^k, \mathbf{p}_{i,2}^k, ..., \mathbf{p}_{i,d}^k\}$ where each decision variable of the optimization problem corresponds to a dimension. After the particle's random initialization within the search space, each individual $\mathbf{p}_i^k$ is evaluated by the use of a fitness function $f(\mathbf{p}_i^k)$, the best fitness value obtained during the evolution process is collected in a vector $\mathbf{g}(g_1, g_2, ..., g_d)$ with their corresponding positions $\mathbf{p}_i^*(\mathbf{p}_{i,1}^*, \mathbf{p}_{i,2}^*, ..., \mathbf{p}_{i,d}^*)$.

For practical proposes, the method used in the study was implemented by Lin in [20] here the PSO calculate the new position for each element as follows:

$$
\begin{aligned}
v_{i,j}^{k+1} &= \omega \cdot v_{i,j}^k + c_1 \cdot r_1 \cdot (p_{i,j}^* - p_{i,j}^k) + c_2 \cdot r_2 \cdot (g_j - p_{i,j}^k), \\
p_{i,j}^{k+1} &= p_{i,j}^k + v_{i,j}^{k+1}
\end{aligned}
\tag{2.1}
$$

where ω is an inertia weight, c_2 and c_1 regulates the updated and the current velocity, and are positive coefficients that control the acceleration velocity of each element towards the positions $\mathbf{g}$ and $\mathbf{p}_i^k$, respectively. r_1 and r_2 are uniformly distributed random numbers within the interval $[0, 1]$.

2.2.2   Artificial Bee Colony (ABC)

The ABC was introduced by Karaboga [14], it is based on the intelligent foraging performance of the honeybee swarm. From the computational point of view, the ABC algorithm produces a population $\mathbf{X}^k (\{\mathbf{x}_1^k, \mathbf{x}_2^k, ..., \mathbf{x}_N^k\})$ of N potential food positions, which are evolved from the initial point ($k = 0$) to stop criterion. Each possible food location $\mathbf{X}_i^k (i \in [1, ..., N])$ symbolizes a d-dimensional vector $\{\mathbf{x}_{i,1}^k, \mathbf{x}_{i,2}^k, ..., \mathbf{x}_{i,d}^k\}$ where the decision variables of the optimization problem correspond to each vector dimension. The first feasible food locations (solutions) are weighed by using an objective function to estimate which food location is an acceptable solution (nectaramount). Once the values from an objective function are obtained, the candidate solution $\mathbf{x}_i^k$ is modified through the ABC operators (honeybee types), where each food location $\mathbf{x}_i^k$ produces a new food source $\mathbf{t}_i$ and is calculated as below:

$$
\mathbf{t}_i = \mathbf{x}_i^k + \phi(\mathbf{x}_i^k - \mathbf{x}_r^k), i, r \in (1, 2, \ldots N);
\tag{2.2}
$$

where $\mathbf{x}_i^k$ is a food location randomly chosen, which obeys the following inequality $r \neq i$. ϕ is a random factor between $[-1, 1]$. After a new element $\mathbf{t}_i$ is produced, a probability correlated with the quality (fitness value) of the solution fit($\mathbf{t}_i$) is determined. With the Eq. (2.4) the fitness value is designated to the candidate solution $\mathbf{x}_i^k$.

$$
\text{fit}(\mathbf{x}_i^k) =
\begin{cases}
\frac{1}{1+f(\mathbf{x}_i^k)} & \text{if } f(\mathbf{x}_i^k) \geq 0 \\
1 + |f(\mathbf{x}_i^k)| & \text{if } f(\mathbf{x}_i^k) < 0
\end{cases};
\tag{2.3}
$$

where $f(\cdot)$ is the objective function of the problem to be treated. After the fitness values are estimated, a greedy selection is employed between $\mathbf{t}_i$ and $\mathbf{x}_i^k$. If fit($\mathbf{t}_i$) is better than fit($\mathbf{x}_i^k$), the candidate solution $\mathbf{x}_i^k$ is replaced by $\mathbf{t}_i$; otherwise, $\mathbf{x}_i^k$ remains.

2.2.3 Differential Evolution (DE)

The DE is a technique developed by Storn and Price [13] based on a parallel direct search, the feasible solutions are represented as a vector and are described as follow:

$$\mathbf{x}_i^t = (\mathbf{x}_{1,i}^t, \mathbf{x}_{2,i}^t, \ldots, \mathbf{x}_{d,i}^t), \quad i = 1, 2, \ldots n; \tag{2.4}$$

where $\mathbf{x}_i^t$ is the ith element at the iteration t. The conventional DE technique has three principal steps: *(a) crossover, (b) mutation,* and *(c) selection* which are detailed below:

(A) *Crossover.* This step is controlled by a crossover rate $C_r \in [0, 1]$. The C_r determines the modification of some elements of the population and it is employed on each dimension of the solution. The crossover uses a uniformly distributed random number $r_i \in [0, 1]$ for determining the jth component $\mathbf{v}_i$ which is manipulated as follows:

$$\mathbf{u}_{j,i}^{t+1} = \begin{cases} \mathbf{v}_{j,i} & r_i \leq C_r \\ \mathbf{x}_{j,i}^t & \text{otherwise} \end{cases}, j = 1, 2, \ldots, d \quad i = 1, 2, \ldots, n; \tag{2.5}$$

(B) *Mutation.* This process is used on a given vector $\mathbf{x}_i^t$ at the t iteration. After three different elements of the population are randomly selected: $\mathbf{x}_p$, $\mathbf{x}_q$ and $\mathbf{x}_r$, a new candidate solution is generated by the following equation:

$$\mathbf{v}_i^{t+1} = \mathbf{x}_p^t + F \cdot (\mathbf{x}_q^t - \mathbf{x}_r^t); \tag{2.6}$$

(C) *Selection.* The selection process is basically the same that is used in any optimization process, the quality of the new solution is evaluated in the fitness function, if the new solution is better than the actual, then it is replaced. To determine the selection criteria, we can consider the following expression:

$$\mathbf{x}_i^{t+1} = \begin{cases} \mathbf{u}_i^{t+1} & \text{if } f(\mathbf{u}_i^{t+1}) \leq f(\mathbf{x}_i^t) \\ \mathbf{x}_i^t & \text{otherwise} \end{cases}; \tag{2.7}$$

2.2.4 Harmony Search (HS)

The HS was introduced by Geem [15], this technique is inspired by music. It was based on the perception that the meaning of music is to search for a perfect state of harmony by improvisation. The analogy in the optimization process is to search for the best value of a certain problem. When a musician improvises, he follows three rules for obtaining a good melody: plays a popular piece of music from his memory, plays something alike to a known piece, and composes a new one, or just plays arbitrary notes. In the HS, those rules become in three essential points; *(a) Harmony Memory (HM) consideration, (b) pitch adjustment,* and *(c) initialize a random solution.*

(a) Harmony Memory (HM) Consideration. This step will guarantee that the best harmonies will be transferred over to the new harmony memory. To handle the HM effectively it is designated a parameter that selects the elements of HM. This parameter is the Harmony Memory Consideration Rate (*HMCR*) and is described as $HMCR \in [0, 1]$. If this value is high, almost all the harmonies are employed in HM, and then other harmonies are not considered in the exploration stage. If the rate of the *HMCR* is low, just a few best harmonies are chosen, and it may converge too slowly.

(b) Pitch Adjustment. Each element achieved by memory consideration is additionally examined to define whether it should be pitch-adjusted. For this operation, the Pitch-Adjusting Rate (*PAR*) is defined to select the frequency of the adjustment and the Bandwidth factor (*BW*) for controlling the local search nearby the chosen elements of the HM. Therefore, the pitch balancing decision is computed as follows:

$$x_{new} = \begin{cases} x_{new} \leftarrow x_{new} \pm \text{rand}(0,1) \cdot BW & \text{with probability } PAR \\ x_{new} & \text{with probability } (1 - PAR) \end{cases}; \quad (2.8)$$

Pitch adjustment is responsible for producing new potential harmonies by slightly adjusting original variable positions. This process can be considered similar to the mutation method in evolutionary algorithms. Hence, the decision variable is either modified by a random number between 0 and *BW* or left unchanged.

(c) Initialize a random solution. This step tries to enhance the diversity of the population. As is specified in (a) if a new solution is not considered from the harmony memory it must be generated a new one in the search space. This process is defined as follows:

$$x_{new} = \begin{cases} x_i \in \{x_1, x_2, \ldots, x_{HMS}\} & \text{with probability } HMCR \\ l + (u - l \cdot \text{rand}(0, 1) & \text{with probability } (1 - HMCR) \end{cases}; \quad (2.9)$$

where *HMS* represents the size of the HM, x_i is an element taken from the HM, and *l* and *u* are the lower and upper bounds in the search space. Finally, rand is a random value uniformly distributed between 0 and 1.

2.2.5   Gravitational Search Algorithm (GSA)

Rashedi proposed the GSA in 2009, this evolutionary algorithm is based on the gravitational laws. In the GSA technique each mass (solution) is represented as a d-dimensional vector; $\mathbf{x}_i(t) = (x_i^1, \ldots, x_i^d)$; $(i \in 1, \ldots, N)$ at a time t, the acting force from a mass i to a mass j of the h variable is described as below:

$$F_{ij}^h(t) = G(t) \cdot \frac{Mp_i(t) \cdot Ma_j(t)}{R_{ij}(t) + \varepsilon} \cdot (x_j^h(t) - x_i^h(t)); \tag{2.10}$$

where Ma_j is the active gravitational mass associated to solution j, Mp_i expresses the passive gravitational mass of solution i, $G(t)$ is the gravitational constant at time t, ε is a small constant, and R_{ij} is the Euclidian distance between the ith and jth solutions. The $G(t)$ function is regulated through the evolution process to regulate the balance between exploration and exploitation by the adjustment of the attraction forces between particles. The total force acting over a certain solution i is described as follow:

$$F_i^h(t) = \sum_{j=1, j \neq i}^{N} F_{ij}^h(t); \tag{2.11}$$

Next, the acceleration of the candidate solution at time t is estimated as follows:

$$a_i^h(t) = \frac{F_i^h(t)}{Mn_i(t)}; \tag{2.12}$$

where Mn_i describes the inertial mass of the candidate solution i. The new position of each solution i is determined as below:

$$\begin{aligned} x_i^h(t+1) &= x_i^h(t) + v_i^h(t+1) \\ v_i^h(t+1) &= rand(\cdot) \cdot v_i^h(t) + a_i^h(t); \end{aligned} \tag{2.13}$$

The gravitational and inertia masses of each solution are estimated in terms of their quality. Hence, the gravitational and inertia masses are updated by the following expression:

$$Ma_i = Mp_i = M_{ii} = M_i \tag{2.14}$$

$$m_i(t) = \frac{f(\mathbf{x}_i(t)) - worst(t)}{best(t) - worst(t)} \tag{2.15}$$

$$M_i(t) = \frac{m_i(t)}{\sum_{j=1}^{N} m_j(t)}, \tag{2.16}$$

where $f(\cdot)$ represents the objective function, whose result determines the quality of solutions.

2.2.6 Flower Pollination Algorithm (FPA)

The FPA was introduced by Yang [16]. This technique is inspired by the flowering pollination process of plants. In the biological procedure, this method is fundamental for the reproduction of plants and is correlated with pollen transfer. This process is conducted by pollinator agents (birds, insects, etc.). This procedure can be grouped into two classes; abiotic and biotic. Most of the plants use the biotic pollination procedure, in which the employment of external pollinators is essential; this method is known as cross-pollination. On the other hand, the abiotic pollination process is better known as self–pollination, it is not necessary the intervention of any external pollinator. The operations for each rule are described in the next subsections. Contrarily to the biological procedure, the FPA considers just one flower that generates just one pollen gamete. The FPA generates a population $\mathbf{F}^k = \left\{ \mathbf{f}_1^k, \mathbf{f}_2^k, ..., \mathbf{f}_m^k \right\}$ of m candidate solutions (flowers) where each solution is represented by a d-dimensional vector $\left\{ f_{i,1}^k, f_{i,2}^k, ..., f_{i,d}^k \right\}$. The number of the parameters of the problem corresponds to the dimensionality of the vector. The algorithm functioning can be summarized is three main steps; *(A) Lévy flights, (B) Local pollination,* and *(C) Global pollination* which are described below:

(A) Lévy flights. Probably the most important operator in the FPA is the Lévy Flights [25, 26], which generates a random walk. The step lengths during the exploration stage are defined by a heavy-tailed probability distribution. The random step is produced as below:

$$s_i = \frac{\mathbf{u}}{|\mathbf{v}|^{1/\beta}} \tag{2.17}$$

where $\mathbf{u} = \{u_1, ..., u_d\}$ and $\mathbf{v} = \{v_1, ..., v_d\}$ are d-dimensional vectors and $\beta = 3/2$. The values of the vectors $\mathbf{u}$ and $\mathbf{v}$ are computed by normal distributions using Eq. (2.18).

$$u \sim N(0, \sigma_u^2), \qquad v \sim N(0, \sigma_v^2),$$
$$\sigma_u = \left(\frac{\Gamma(1+\beta) \cdot sin(\pi \cdot \beta/2)}{\Gamma((1+\beta)/2) \cdot \beta \cdot 2^{(\beta-1)/2}} \right)^{1/\beta}, \qquad \sigma_v = 1, \tag{2.18}$$

where $\Gamma(\cdot)$ expresses the Gamma distribution.

(B) Local pollination. The local pollination is defined by Eq. (2.19), where $\mathbf{f}_j^k$ and $\mathbf{f}_h^k$ are pollens from two different flowers of the same population. The next solution is calculated using the following expression:

$$\mathbf{f}_i^{k+1} = \mathbf{f}_i^k + \varepsilon(\mathbf{f}_i^k - \mathbf{f}_h^k) \tag{2.19}$$

where ε is a uniformly random value within $[0, 1]$.

(C) Global pollination. In the global pollination, the next element is determined by the follow equation:

$$\mathbf{f}_i^{k+1} = \mathbf{f}_i^k + s_i(\mathbf{g}^* - \mathbf{f}_i^k) \tag{2.20}$$

where s_i is a uniformly random value within $[0, 1]$ while $\mathbf{g}^*$ is the best solution observed so-far.

2.3 2D-IIR Filter Design Procedure

In this section, the procedure of the 2D-IIR filter design as an optimization problem is formulated. In this chapter, the zero-phase IIR filter is considered. The zero-phase filter is a special example of a linear phase filter where the phase slope is zero. This fact indicates that the real impulse response is always even (it satisfies $h(n) = h(-n)$) [6]. The structure of a 2D-IIR filter employed in the experimental stage is presented in Eq. (2.21) [27–31].

$$H(z_1, z_2) = H_0 \cdot \frac{\sum_{i=0}^{K} \sum_{j=0}^{K} a_{ij} z_1^i z_2^i}{\prod_{k=1}^{K} (1 + b_k z_1 + c_k z_2 + d_k z_1 z_2)} \tag{2.21}$$

where a_{ij}, b_k, c_k, d_k are the filter design coefficients, is an intensity constant while K expresses the filter order. In the case of $K = 2$, the model of Eq. 2.21 corresponds to a 2nd order filter whereas a higher value ($K > 3$) defines a high order filter. Equation 2.21 is usually configured as $a_{00} = 1$ (by the normalization of a_{ij} to the value of a_{00}). z_1 and z_2 describe complex variables in the z-transform domain. Nevertheless, for practicality, the Fourier frequency domain is employed, where z_1 and z_2 are represented as below:

$$z_1 = e^{j\omega_1}, z_2 = e^{j\omega_2} \tag{2.22}$$

where j is an imaginary number and $\omega_1, \omega_2 \in [0, \pi]$ are the frequency variables. With the replacement of Eq. 2.22 in Eq. 2.21, the expression $H(z_1, z_2)$ change to $M(\omega_1, \omega_2)$.

The main objective of a filter design is to approximate the response of a defined 2D-IIR filter structure to the desired behavior $M_d(\omega_1, \omega_2)$ given by the user specifications. Consequently, the filter design process consists of determining the best parameters of $M(\omega_1, \omega_2)$ that represent the best possible correspondence with the desired response $M_d(\omega_1, \omega_2)$. Both responses $M(\omega_1, \omega_2)$ and $M_d(\omega_1, \omega_2)$ are discretely calculated at $N_1 \times N_2$ points.

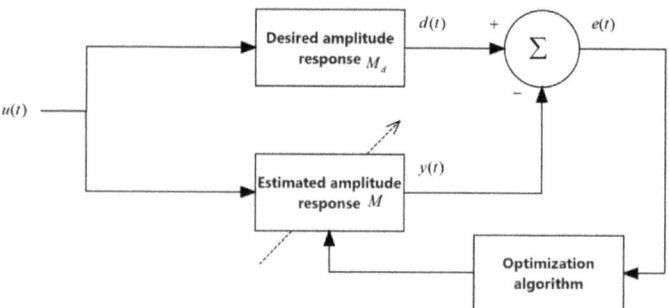

Fig. 2.1 Diagram of 2D IIR filter design based on desired amplitude response

To conduct the search procedure, an objective function J that estimates the similarity between the candidate filter design $M(\omega_1, \omega_2)$ and the desired response $M_d(\omega_1, \omega_2)$ must be determined. This similarity can be represented as the difference between both responses, as is shown in Eq. 2.23.

$$J(a_{ij}, b_k, c_k, d_k, H_0) = \sum_{n_1=0}^{N_1} \sum_{n_2=0}^{N_2} [|M(\omega_1, \omega_2)| - M_d(\omega_1, \omega_2)]^2 \qquad (2.23)$$

where $\omega_1 = (\pi/N_1) \cdot n_1$, $\omega_2 = (\pi/N_2) \cdot n_2$. For purpose of simplicity, Eq. (2.23) is rewritten as:

$$J = \sum_{n_1=0}^{N_1} \sum_{n_2=0}^{N_2} \left[\left| M\left\{ \frac{\pi n_1}{N_1}, \frac{\pi n_2}{N_2} \right\} \right| - M_d\left\{ \frac{\pi n_1}{N_1}, \frac{\pi n_2}{N_2} \right\} \right]^2 \qquad (2.24)$$

Figure 2.1 shows the filter design process. In the computational schema, the difference between the calculated $(y(t))$ and desired $(d(t))$ response is employed to control the optimization process.

The principal challenge of 2D IIR filter design is the trend to generate unstable transfer functions. Hence, before evaluating the Eq. 2.24, it is essential to verify that the poles of the transfer function are positioned within the unitary circle of the z-plane. To verify this state, different conditions must be satisfied, which for Eq. 2.25 can be represented as follows [28, 29]:

$$|b_k + c_k| - 1 < d_k < 1 - |b_k - c_k|, k = 1, 2, \ldots, K \qquad (2.25)$$

Under such considerations, the filter design procedure can be turned into an optimization problem.

2.3.1 Comparative Parameter Setting

To carry out the comparative study, it is necessary to determine the experiment and its main features. In the experiment, it has been chosen the filter design problem studied in [30, 31]. The idea of this selection is to maintain compatibility with the results reported by such studies. In the test, the 2D-IIR filter structure is expressed as follows:

$$H(z_1, z_2) = H_0 \cdot \frac{a_{00} + a_{01}z_2 + a_{02}z_2^2 + a_{10}z_1 + a_{20}z_1^2 + a_{11}z_1z_2 + a_{12}z_1z_2^2 + a_{21}z_1^2z_2 + a_{22}z_1^2z_2^2}{(1 + b_1z_1 + c_1z_2 + d_1z_1z_2)(1 + b_2z_1 + c_2z_2 + d_2z_1z_2)} \quad (2.26)$$

For computation facility, the variables z_1 and z_2 are changed by $f_{xy} = \cos(x\omega_1 + y\omega_2)$ and $g_{xy} = \sin(x\omega_1 + y\omega_2)$, $[x, y] = [0, 1, 2]$. This modification has been introduced by Swagatam and Konar in [27]. The main objective is to make easier the implementation of digital filters design. Hence, Eq. (2.26) is reformulated as:

$$M(\omega_1, \omega_2) = H_0 \left[\frac{a_{00} + a_{01}f_{01} + a_{02}f_{02} + a_{10}f_{10} + a_{20}f_{20} + a_{11}f_{11} + a_{12}f_{12} + a_{21}f_{21} + a_{22}f_{22}}{V} \cdots \right.$$
$$\left. \cdots - j\frac{a_{00} + a_{01}g_{01} + a_{02}g_{02} + a_{10}g_{10} + a_{20}g_{20} + a_{11}g_{11} + a_{12}g_{12} + a_{21}g_{21} + a_{22}g_{22}}{V} \right] \quad (2.27)$$

where V is described by:

$$V = [(1 + b_1 f_{10} + c_1 f_{01} + d_1 f_{11}) - j(b_1 g_{10} + c_1 g_{01} + d_1 g_{11}] \cdot$$
$$[(1 + b_1 f_{10} + c_1 f_{01} + d_1 f_{11}) - j(b_1 g_{10} + c_1 g_{01} + d_1 g_{11}] \quad (2.28)$$

Under such circumstances, Eq. (2.27) can be expressed as follows:

$$M(\omega_1, \omega_2) = H_0 \cdot \frac{N_R - jN_I}{(D_{1R} - jD_{1R})(D_{2R} - jD_{2I})} \quad (2.29)$$

where

$$N_R = a_{00} + a_{01}f_{01} + a_{02}f_{02} + a_{10}f_{10} + a_{20}f_{20} + a_{11}f_{11} + a_{12}f_{12}$$
$$+ a_{21}f_{21} + a_{22}f_{22}$$
$$N_I = a_{00} + a_{01}g_{01} + a_{02}g_{02} + a_{10}g_{10} + a_{20}g_{20} + a_{11}g_{11} + a_{12}g_{12}$$
$$+ a_{21}g_{21} + a_{22}g_{22}$$
$$D_{1R} = 1 + b_1 f_{10} + c_1 f_{01} + d_1 f_{11}$$
$$D_{1I} = 1 + b_1 g_{10} + c_1 g_{01} + d_1 g_{11}$$
$$D_{2R} = 1 + b_2 f_{10} + c_2 f_{01} + d_2 f_{11}$$
$$D_{2I} = 1 + b_2 g_{10} + c_2 g_{01} + d_2 g_{11} \quad (2.30)$$

Considering Eq. (2.29) and Eq. (2.30) the magnitude of the filter $|M(\omega_1, \omega_2)|$ can be calculated as:

$$|M(\omega_1, \omega_2)| = H_0 \sqrt{\frac{N_R^2 - N_I^2}{(D_{1R}^2 - D_{1I}^2)(D_{2R}^2 - D_{2I}^2)}} \tag{2.31}$$

In the experiment, the desired response $M_d(\omega_1, \omega_2)$ is represented as a circular symmetric low-pass response, which is described as follows:

$$M_d(\omega_1, \omega_2) = \begin{cases} 1 & \text{if } \sqrt{\omega_1^2 + \omega_2^2} \leq 0.08\pi \\ 0.5 & \text{if } 0.08\pi \leq \sqrt{\omega_1^2 + \omega_2^2} \leq 0.12\pi \\ 0 & \text{otherwise} \end{cases} \tag{2.32}$$

In order to approximate both responses $M(\omega_1, \omega_2)$ and $M_d(\omega_1, \omega_2)$, it is employed a matrix of $N_1 \times N_2$ points, where N_1 and N_2 are set in 50. To guarantee the stability of the filter, the candidate solution must satisfy the following constraints:

$$\begin{aligned} -(1 + d_k) &< (b_k + c_k) < (1 + d_k) \\ -(1 - d_k) &< (b_k + c_k) < (1 - d_k) \\ (1 + d_k) &> 0 \\ (1 - d_k) &> 0 \end{aligned} \tag{2.33}$$

Under such circumstances, the filter design procedure can be described as a constrained optimization problem. Thie optimization task is defined as follows:

$$\min J(a_{ij}, b_k, c_k, d_k, H_0) = \sum_{n_1=0}^{N_1} \sum_{n_2=0}^{N_2} \left[\left| M\left\{ \frac{\pi n_1}{50}, \frac{\pi n_2}{50} \right\} \right| - M_d\left\{ \frac{\pi n_1}{50}, \frac{\pi n_2}{50} \right\} \right]^2$$

$$\begin{aligned} \text{subject to } -(1 + d_k) &< (b_k + c_k) < (1 + d_k) \\ -(1 - d_k) &< (b_k + c_k) < (1 - d_k) \\ (1 + d_k) &> 0 \qquad\qquad k = 1, 2, \ldots, K \\ (1 - d_k) &> 0, \end{aligned} \tag{2.34}$$

To employ an optimization technique to the problem shown in Eq. (2.34) is necessary to express each candidate solution as one position in the multi-dimensional search space. The candidate solution of Eq. (2.34) considers a 2nd order filter $K = 2$, and is represented by a 15-dimensional vector $\mathbf{x}_i$:

$$\mathbf{x}_i = [a_{01}, a_{02}, a_{10}, a_{11}, a_{12}, a_{20}, a_{21}, a_{22}, b_1, c_1, d_1, b_2, c_2, d_2, H_0]^T \tag{2.35}$$

The optimal solution is expressed by the solution $\mathbf{x}_i$ that minimizes Eq. (2.34), while it simultaneously holds the conditions required by its corresponding constraints. The best solution corresponds to the 2D-IIR filter whose response $M(\omega_1, \omega_2)$ is the most similar to $M_d(\omega_1, \omega_2)$.

For a constraint approach, all optimization technique includes a penalty factor [32] whose value corresponds to the violated restrictions. The purpose is to include a tendency term into the objective function to penalize constraint violations for avoiding unfeasible solutions.

2.4 Experimental Results

In this section, a comparative study is presented. In the experimentation, the 2D-IIR filter design problem formulated in Sect. 2.3 is used. In the experimentation, six different algorithms were used: the Swarm Optimization (PSO) [12], Differential Evolution (DE) [13], Artificial Bee Colony (ABC) algorithm [14], Harmony Search (HS) [15], Flower Pollination Algorithm (FPA) [16], and Gravitational Search Algorithm (GSA) [17]. These techniques are considered the most popular optimization techniques currently in use [11]. For the comparative study, the parameters of all techniques are set as follows:

1. PSO, $c_1 = 2$, $c_2 = 2$ and weights factors were set $w_{max} = 0.9$, and $w_{min} = 0.4$ [33].
2. ABC, all parameters were provided by [14], *limit = 90*.
3. HS, agreeing with [15], the parameters were set $HMCR = 0.95$, and $PAR = 0.3$.
4. DE, in concordance with [13] the parameters were set $CR = 0.9$ and $f = 0.5$.
5. GSA, $G_0 = 100$, and $\alpha = 20$, according to [17].
6. FPA, the probabilistic global pollination factor was set $p = 0.8$ [16].

All optimization techniques have been configured with a population of 25 individuals, while the stop criteria is set to 3000 iterations. In this work, the generated filters have not been implemented in hardware, for this reason, the analysis of the quantization error is not considered. All experimental results have been obtained through simulation.

2.4.1 Accuracy Comparison

For the experiment, a 2nd order filter design from Eqs. (2.34) and (2.32) is considered. Table 2.1 shows the J values produced by the experimentation. They estimate the differences between the ideal response and those generated by the produced filters through optimization techniques. A lower value of J suggests a better approximation of the desired response. In the table, the minimal (J_{min}), maximal (J_{max}) and mean ($\overline{J}$) values of J along with their standard deviations (σ) obtained for each technique during 250 executions are reported. The first three indexes reflect the accuracy of each optimization technique, whereas the last one represents its robustness. The experimental results show that the ABC and FPA techniques show the best performance considering the quality indexes employed for the test. Moreover, the ABC technique has the lowest standard deviation values, which means that it can determine the same solution with a very low variation. The rest of the optimization techniques achieve different performance levels. In Table 2.1, the highlighted values describe the best-found values.

Table 2.2 shows the parameters of the best-found filter achieved by each optimization technique, PSO, ABC, DE, HS, GSA, and FPA. Such values correspond to the parameters that generate minimal J value (J_{min}). In Table 2.2, the highlighted values describe the best-found parameters among all algorithms.

In Fig. 2.2 the filter responses generated by each technique are presented. From graphic b to g, the filter responses produced according to the parameters exhibited in Table 2.2 are presented. After a quick examination of Fig. 2.2, it is clear that PSO and HS techniques produce the worst results since raised ripples can be appreciated in their responses. GSA and DE present lower ripples than PSO and HS. On the other

Table 2.1 Minimum (J_{min}) and maximum (J_{max}) values of the objective function, Mean ($\overline{J}$) and Standard deviation (σ) over 250 executions for each technique

	PSO	ABC	DE	HS	GSA	FPA
J_{min}	2.6971	3.0655	2.6550	3.4276	3.2668	**2.3030**
J_{max}	16.1331	4.0599	15.1968	8.5494	16.9647	**3.3276**
$\overline{J}$	6.5976	3.5173	5.2582	4.5836	6.3707	**2.5838**
σ	3.4537	**0.1736**	2.0377	0.5925	4.1108	0.2759

Table 2.2 2D-IIR filter coefficients achieved by PSO, ABC, DE, HS, GSA and FPA techniques

2D-IIR filter coefficients	PSO	ABC	DE	HS	GSA	FPA
a_{01}	0.1182	0.3217	−0.4785	−0.7521	0.2978	**−0.0388**
a_{02}	0.0149	0.3109	−0.9983	−0.2398	−0.0654	**−0.6939**
a_{10}	0.6393	−0.5132	−1.0000	−0.3936	−0.6860	**−0.5417**
a_{11}	0.2938	−0.9392	−0.8839	0.1290	0.6559	**0.0738**
a_{12}	0.3514	−0.1289	0.9422	−0.3705	−0.8044	**−0.4146**
a_{20}	0.3459	0.7601	0.7540	−0.9324	−0.1292	**−0.2994**
a_{21}	0.9404	−0.0376	−0.8596	0.1805	0.3098	**−0.8339**
a_{22}	−0.9137	0.8856	−0.4131	0.2330	0.2307	**0.6927**
b_1	−0.2290	0.3588	−0.9353	−0.3858	−0.2743	**−0.2251**
b_2	−0.9081	−0.9238	0.5207	−0.6663	−0.8617	**−0.9241**
c_1	−0.8491	−1.0000	−0.9046	−0.3996	−0.8175	**−0.1543**
c_2	−0.7313	−0.6050	−0.8622	−0.5995	−0.5501	**−0.9438**
d_1	0.1360	−0.4827	0.8458	−0.2974	0.1367	**−0.8157**
d_2	0.6545	0.5521	−0.9204	0.2159	0.4369	**0.8906**
H_0	0.0004	0.0019	0.0009	0.0046	0.0013	**0.0038**

hand, ABC and FPA achieve the best filter responses. The ABC determine a response where the ripples are small whereas the FPA response practically eliminates them.

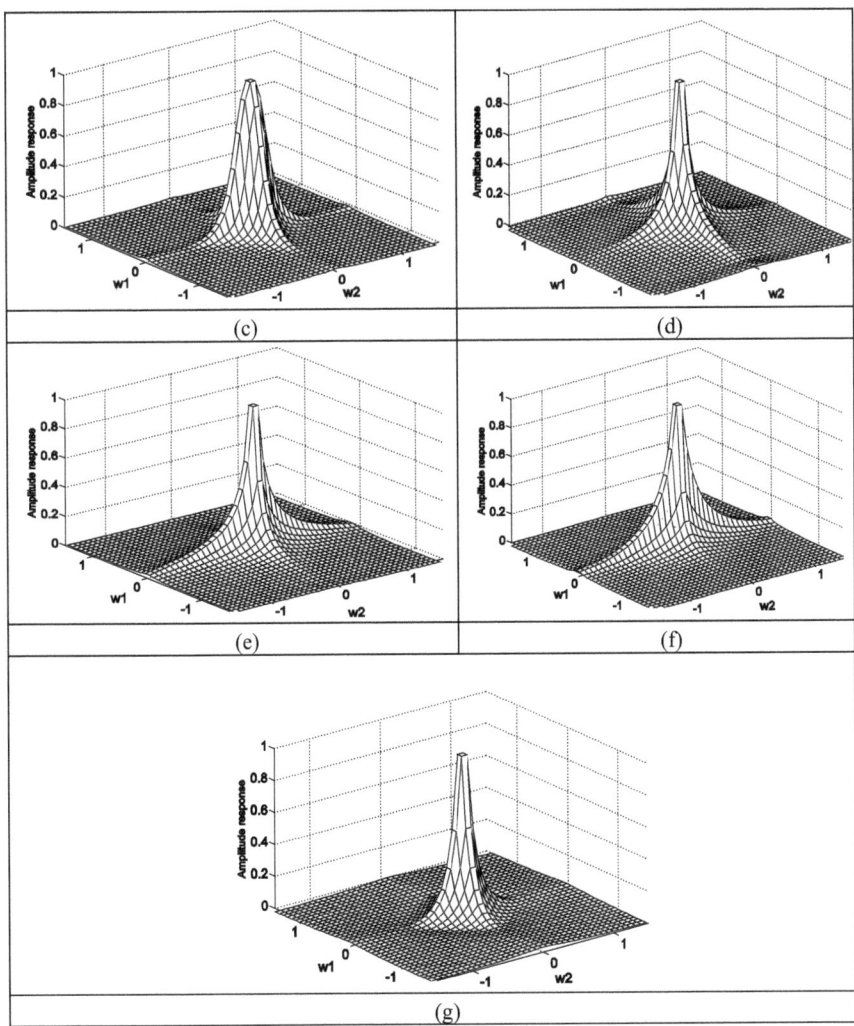

Fig. 2.2 **a** Desired amplitude response, **b** amplitude response determined by PSO, **c** amplitude response determined by ABC, **d** amplitude response determined by DE, **e** amplitude response determined by HS, **f** amplitude response determined by GSA, **g** amplitude response determined by FPA

2.4.2   Convergence Study

In the comparative study of $\overline{J}$ values, cannot be completely described the searching performance of the optimization techniques. Therefore, in Fig. 2.3, a convergence analysis considering the six optimization techniques has been carried out. The main idea of this test is to estimate the speed of the techniques for reaching the optimal values. For this experiment, the performance of each technique considering the minimization of the fitness function for the filter design is analyzed. Figure 2.3 shows the median result after 250 individual executions. In the figure, it is clear that PSO and GSA converge faster than the other techniques, but the premature convergence causes that the solutions get stuck in local minima.

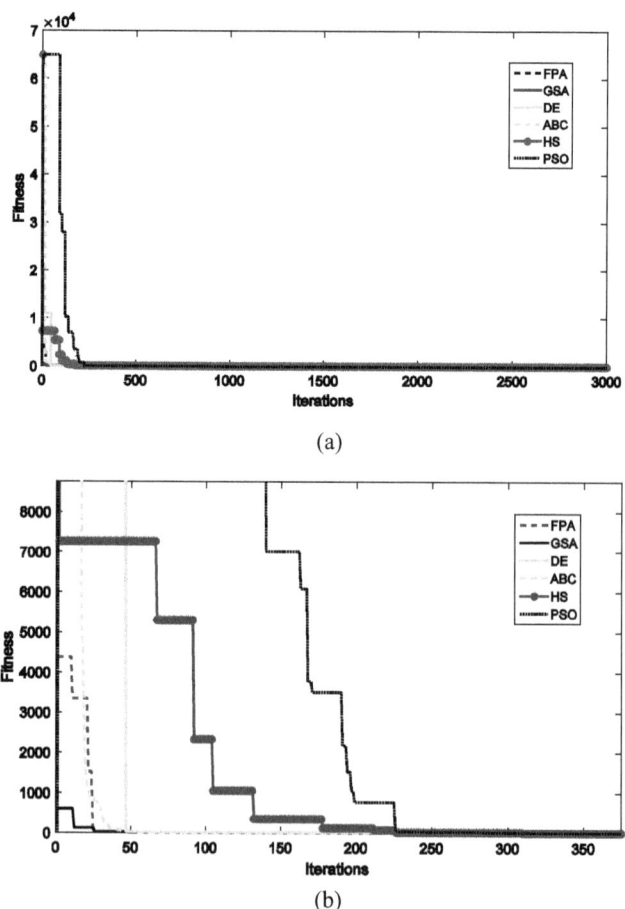

Fig. 2.3 The evolution process of the objective function for FPA, GSA, DE, HS, ABC, and PSO. **a** Graphs over 3000 iterations and **b** zoom over 400 iterations

Table 2.3 Average measures of function evaluations (*NFE*) and the computational time (*CT*) spent by each technique collected from 20 executions

Index	PSO	ABC	DE	HS	GSA	FPA
NFE	48,543.12	58,321.09	57,322.24	49,322.74	56,654.57	51,342.18
CT (s)	8.25	12.54	13.02	9.10	11.82	10.12

2.4.3 Computational Cost

In the experimental test, the main objective is to estimate the number of function evaluations (*NFE*) and the computing time (*CT*) employed by each technique required to determine the parameters of the filter design. All tests consider the filter design expressed by Eqs. 2.32 and 2.34 which agree to the parameters shown in Table 2.2. The *NFE* index indicates the number of function evaluations from which improvement is not necessarily identified in the objective function (Eq. 2.34). CT denotes the passed time in seconds spent by each technique corresponding to the *NFE*. The experimentation has been performed in Matlab operating in a Computer Nitro, Core i7. Table 2.3 exhibits the averaged measures which are collected from 20 independent executions. A close examination of Table 2.3 proves that PSO and HS are the fastest techniques. Although, this apparent velocity corresponds to premature convergence. ABC, DE, and GSA are the slowest approaches. On the other hand, FPA is the scheme that obtains the best trade-off between speed and accuracy.

2.4.4 Comparison with Different Bandwidth Sizes

To continue with the comparative analysis, the performance of each technique in the low pass filter design with different bandwidth sizes is evaluated. For this test, two special cases are considered. In the first case, the aim is to design a filter with an ideal response determined by:

$$M_d^1(\omega_1, \omega_2) = \begin{cases} 1 & \text{if } \sqrt{\omega_1^2 + \omega_2^2} \leq 0.16\pi \\ 0.5 & \text{if } 0.16\pi \leq \sqrt{\omega_1^2 + \omega_2^2} \leq 0.24\pi \\ 0 & \text{otherwise} \end{cases} \quad (2.36)$$

The second case examines the design of a filter with the following ideal response:

$$M_d^2(\omega_1, \omega_2) = \begin{cases} 1 & \text{if } \sqrt{\omega_1^2 + \omega_2^2} \leq 0.32\pi \\ 0.5 & \text{if } 0.32\pi \leq \sqrt{\omega_1^2 + \omega_2^2} \leq 0.48\pi \\ 0 & \text{otherwise} \end{cases} \quad (2.37)$$

Table 2.4 Minimum (J_{min}^p) and maximum (J_{max}^p) results of the objective function, Mean ($\overline{J}^p$) and Standard deviation (σ_p) for the results for each case $p = 1, 2$

	PSO	ABC	DE	HS	GSA	FPA
J_{min}^1	15.94	13.01	20.93	23.12	22.67	12.68
J_{min}^2	36.54	35.72	18.90	24.33	58.46	17.52
J_{max}^1	73.68	16.85	35.84	47.58	87.91	15.24
J_{max}^2	155.63	51.91	31.39	41.48	45.15	26.35
$\overline{J}^1$	25.40	14.94	25.21	31.19	75.87	13.95
$\overline{J}^2$	68.62	44.45	24.65	28.94	49.94	22.75
σ_1	14.17	0.99	3.14	6.90	17.60	1.54
σ_2	28.81	4.85	3.44	3.75	3.13	3.86

Considering J^1 and J^2 as objective functions agreeing to each ideal response M_d^1 and M_d^2, in Table 2.4 the comparison results are shown. They determine the differences between the response generated by the designed filters through EC techniques for different bandwidth sizes and the ideal filter responses. The minimal (J_{min}^p), maximal (J_{max}^p) and mean ($\overline{J}^p$) values of J along with their standard deviations (σ_p) generated for each technique for the cases $p = 1,2$ are presented. The achieved values are computed from 250 executions of each algorithm. Table 2.4 shows that the FPA technique provides the best performance for both examples J^1 and J^2.

2.4.5   Filter Performance Features

The performance of the filter properties is analyzed in this subsection. The aim is to estimate the responses to the best filters generated by each technique PSO, ABC, DE, HS, GSA, and FPA (reported in Table 2.5). The responses are evaluated considering the next indexes: (a) error in the pass-band, (b) magnitude ripple, (c) stop-band attenuation and (d) the magnitude of poles.

Table 2.5 Performance of the filter components. The results are the responses for each of the best filters generated by every technique PSO, ABC, DE, HS, GSA, and FPA

Index	PSO	ABC	DE	HS	GSA	FPA
$E_{RP}(\omega)$	0.0881	0.0097	0.0631	0.0287	0.0714	0.0051
$E_{MR}(\omega)$	0.1421	0.0918	0.1977	0.1021	0.1391	0.0272
$E_{MR}(\omega)$	25.87	34.21	26.42	35.21	29.28	42.18
MPs	0.8721	0.9180	0.9182	0.8916	0.9019	0.8460

(a) Error in the pass-band.

It estimates the error in the frequency in the pass-band considering the following expression:

$$E_{RP}(\omega) = \left|M(e^{j\omega}) - M_d(e^{j\omega})\right| \text{ for } \omega \in [0, \omega_p], \tag{2.38}$$

where is to the pass-band frequency.

(b) Ripple magnitude in pass-band.

It studies the ripple size considering the equation below:

$$E_{MR}(\omega) = \left|M(e^{j\omega})\right| - 1 \text{ for } \omega \in [0, \omega_p]. \tag{2.39}$$

(c) Attenuation in the stop-band.

It defines the response debilitation in the stop-band as below:

$$AT(\omega) = 20\log_{10}\left(\left|M(e^{j\omega})\right|\right) \text{ for } \omega \in [\omega_s, \pi], \tag{2.40}$$

where is the stop-band frequency.

(d) Maximal pole size (MPs).

This measure estimates the maximal magnitude of poles in $M(e^{j\omega})$.

The error generated by the frequency response in the pass-band $E_{RP}(\omega)$ is the difference between the filter design $M(e^{j\omega})$ and the desired filter response $M_d(e^{j\omega})$. Under such circumstances, a small pass-band error as possible describes a better filter response. On the other hand, the ripple magnitude in pass-band $E_{MR}(\omega)$ has to keep small enough to ensure a lower distortion in the pass-band response. Contrarily, the attenuation $E_{MR}(\omega)$ should be large with the purpose to reduce undesired elements of high frequency. Table 2.5 shows the performance comparison of the filter components. The results are the responses of each filter produced by every technique PSO, ABC, DE, HS, GSA, and FPA. Table 2.5 demonstrate that the FPA algorithm maintains the best performance regarding the error in pass-band, magnitude ripple, stop-band attenuation, and magnitude of poles.

2.4.6 Statistical Non-parametrical Analysis

To statistically verify the results generated by all techniques, a non-parametric statistical significance analysis is implemented. This statistical analysis is known as Wilcoxon's rank-sum test [22, 24, 34]. To use this study, 35 executions of each technique were performed to prove that there is a significant difference between the values provided by each algorithm. The analysis is conducted considering a 5%

significance level (0.05) over the data. Once the ABC and FPA are the methods that better perform the filter design, we continue to compare them with the rest of the techniques. In Table 2.6, the p-values generated by Wilcoxon's among FPA versus PSO, FPA versus ABC, FPA versus DE, FPA versus HS, FPA versus GSA. Table 2.7 describes the results between ABC versus PSO, ABC versus DE, ABC versus HS, ABC versus FPA. In the case of PSO, DE, HS, and GSA, Wilcoxon's test was not conducted, since the obtained results are the worst of all techniques. In the non-parametric study, is considered as a null hypothesis that there is no important difference between the two techniques. On the other hand, it is accepted as an alternative hypothesis that the difference between the two approaches is statistically enough. To aid the interpretation of Table 2.6, the symbols ▲, ▼, and ▶ are chosen. ▲ symbolizes that the proposed technique performs statistically better than the algorithm used on the specified function. ▼ expresses that the technique performs worse than the compared algorithm, and ▶ means that the Wilcoxon test is not able to differentiate between both techniques. Tables 2.6 and 2.7 show that all p-values in the columns are within the range [0, 0.05], which symbolizes that the FPA performs better than the rest of the techniques.

Since we have performed Wilcoxon's tests on several occasions, the possibility of committing an error type 1 (false positive) increases, under such circumstances, we can discard the null hypothesis even when it is true. To discard this issue, a modification for the p-values of the Wilcoxon test is carried out by employing the Bonferroni correction [35]. Using the Wilcoxon test and the Bonferroni correction, we can determine that exists a statistical difference between the techniques used for the study. The Bonferroni correction results for FPA and ABC are presented in Tables 2.8 and 2.9, respectively. The new significance value generated by Bonferroni was set to $p = 0.001428$ for the FPA comparison, and for the ABC comparison, the value has been set on $p = 0.000084034$. In this test, if the new Bonferroni value is marked as 1, it indicates that this element verifies the result of the Wilcoxon study for a particular case.

Table 2.6 p-values of the Wilcoxon's test for FPA versus PSO, ABC, DE, HS, and GSA

FPA versus	PSO	ABC	DE	HS	GSA
Wilcoxon test	1.41e−12▲	1.81e−12▲	9.22e−13▲	6.54e−13▲	7.13e−13▲

Table 2.7 p-values of the Wilcoxon's test for ABC versus PSO, DE, HS, GSA and FPA

ABC versus	PSO	DE	HS	GSA	FPA
Wilcoxon test	4.07e−11▲	3.45e−09▲	1.53e−12▲	2.62e−06▲	1.81e−12▼

Table 2.8 Bonferroni test for FPA versus all techniques

FPA versus	PSO	ABC	DE	HS	GSA
$p <=$ Bonferroni	1	1	1	1	1

Table 2.9 Bonferroni test for ABC versus all techniques

ABC versus	PSO	DE	HS	GSA	FPA
$p <=$ Bonferroni	1	1	1	1	1

2.4.7 Filter Design Study in Images

Figures 2.4 and 2.5 show the performance of the designed filters employed on images for eliminating Gaussian noise for two different images. Figure 2.4 presents the noise removing results of the filter design presented in the last study. The filters have been designed using the PSO, ABC, DE, HS, GSA, and FPA regarding the approximation of a low pass response. The results confirm the capability of each filter for removing noise. Figure 2.4a displays the original image "Lenna". Figure 2.4b presents the image with Gaussian noise of mean $= 0$ and variance of 0.005. The images from Fig. 2.4c–h evidence the performance of each technique PSO, ABC, DE, HS, GSA, and FPA for the noise removing process. It is clear that the filters generated by the FPA, and ABC techniques have the best performance. On the other hand, the filters produced by the PSO, DE, HS, and GSA determine the worst filtering effect. They cannot remove appropriately the added noise, producing distortions in the image due to the frequency ripples [36–38]

Figure 2.5 presents the noise-removing results with a wider bandwidth size filter. As in the last example, they correspond to the filters provided by the PSO, ABC, DE, HS, GSA, and FPA techniques considering the ideal response defined by M_d^1. Figure 2.5a shows the image "Cameraman". Figure 2.5b displays the noisy image corrupted with Gaussian of mean $= 0$ and variance of 0.005. Images from Fig. 2.5c–h present the performance of each technique, the PSO, ABC, DE, HS, GSA, and FPA in the noise-removing process. Moreover, the filters generated by the FPA, and ABC techniques perform the best noise removal result. Contrarily, the filters designed by the PSO, DE, HS, and GSA provide a bad filtering effect.

2.5 Conclusions

Two-dimensional Infinite Impulse Response (2D-IIR) filters have recently attracted attention in many fields of engineering due to their wide range of applications. Incorporating a user-defined filter in a 2D-IIR structure can be described as an optimization problem. On the other hand, Evolutionary Computation (EC) techniques are techniques with high capabilities to explore complex search spaces for a suitable solution. In this chapter, the comparative study of various EC algorithms when they are employed in the 2D-IIR filter design is reported. In the comparison, the most popular EC techniques currently in use such as PSO, ABC, DE, HS, and GSA are considered. In the experimentation, the statistical analysis of Wilcoxon and Bonferroni have been performed in order to validate the results achieved by the techniques.

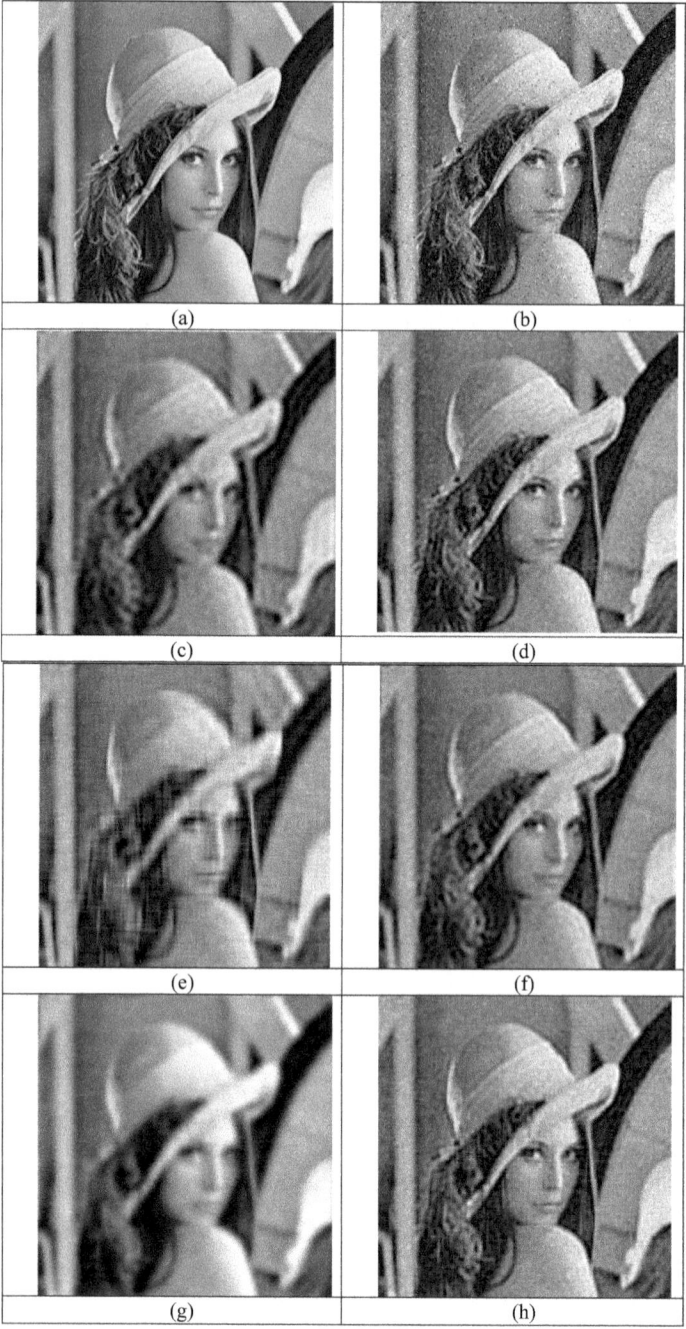

Fig. 2.4   A visual study of designed filters. **a** Original image, **b** noisy image, **c** filtering with PSO design, **d** filtering with ABC design, **e** filtering with DE design, **f** filtering with HS design, **g** filtering with GSA design, and **h** filtering with FPA design

Fig. 2.5 A visual study of designed filters considering a wider bandwidth. **a** Original image, **b** noisy image, **c** filtering with PSO design, **d** filtering with ABC design, **e** filtering with DE design, **f** filtering with HS design, **g** filtering with GSA design, and **h** filtering with FPA design

In the comparison, a 2D-IIR digital filter widely used in literature has been evaluated, in which the optimization of its cost function is particularly complicated, due to the requirements for its design. The results achieved by FPA outperform the PSO, DE, HS, and GSA regarding the average and standard deviation of the fitness values over 250 executions. While the ABC overcomes the FPA in standard deviation, which indicates that ABC can converge around the same solution with a minimal variation, it indicates that FPA and ABC are the best alternatives for this specific application.

As part of the experimentation, a non-parametric analysis known as Wilcoxon's test was conducted. In this analysis, we examined the techniques that better performed the 2D-IIR filter design, the non-parametric test showed that there is a statistical difference between the results archived by FPA and the other techniques employed for the comparison, proving its consistency and accuracy for the 2D-IIR filter design.

The ABC was considered in the Wilcoxon analysis due to the fitness values were consistent, generating the lowest standard deviation in all experiments. In order to avoid error type 1, we applied Bonferroni correction, which showed that the null hypothesis was rejected, with this, we have evidence to guarantee that FPA outperforms the techniques employed for 2D-IIR digital filter design. The notable performance of FPA is its good balance between exploration and exploitation stages and its powerful operators such as Lévy flights, which allow better accuracy in its solutions.

References

1. Srivastava VK, Ray GC (2000) Design of 2D-multiple notch filter and its application in reducing blocking artifact from DCT coded image. 2829–2833
2. Rashedi E, Zarezadeh A (2014) Noise filtering in ultrasound images using gravitational search algorithm. Iran Conf Intell Syst 2014:1–4. https://doi.org/10.1109/IranianCIS.2014.6802559
3. Sgro JA, Emerson RG, Pedley TA (1985) Real-time reconstruction of evoked potentials using a new two-dimensional filter method. Electroencephalogr Clin Neurophysiol Potentials Sect 62:372–380. https://doi.org/10.1016/0168-5597(85)90046-2
4. Dumitrescu B (2005) Optimization of two-dimensional IIR filters with nonseparable and separable denominator. IEEE Trans Signal Process 53:1768–1777. https://doi.org/10.1109/TSP.2005.845469
5. Das S, Konar A (2006) Two-dimensional IIR filter design with modern search heuristics: a comparative study. Int J Comput Intell Appl 06:329–355. https://doi.org/10.1142/s1469026806001848
6. Lu W-S, Andreas A (1992) Two-dimensional digital filters. CRC PressI Llc, New York
7. Dhabal S, Venkateswaran P (2014) Two-dimensional IIR filter design using simulated annealing based particle swarm optimization. J Optim 2014:239721. https://doi.org/10.1155/2014/239721
8. Ampazis N, Perantonis SJ (2013) An efficient constrained learning algorithm for stable 2D IIR filter factorization. Adv Artif Neural Syst 2013:1–7. https://doi.org/10.1155/2013/292567
9. Jin Y, Branke J (2005) Evolutionary optimization in uncertain environments—a survey. IEEE Trans Evol Comput 9:303–317. https://doi.org/10.1109/TEVC.2005.846356
10. Kukrer O (2011) Analysis of the dynamics of a memoryless nonlinear gradient IIR adaptive notch filter. Signal Process 91:2379–2394. https://doi.org/10.1016/j.sigpro.2011.05.001

11. Nanda SJ, Panda G (2014) A survey on nature inspired metaheuristic algorithms for partitional clustering. Swarm Evol Comput 16:1–18. https://doi.org/10.1016/J.SWEVO.2013.11.003
12. Kennedy J, Eberhart R (1995) Particle swarm optimization. IEEE Int Conf Neural Netw 4:1942–1948. https://doi.org/10.1109/ICNN.1995.488968
13. Storn R, Price K (1997) Differential evolution—a simple and efficient heuristic for global optimization over continuous spaces. J Glob Optim 341–359. https://doi.org/10.1023/A:100 8202821328
14. Karaboga D (2005) An idea based on honey bee swarm for numerical optimization. Tech Rep TR06, Erciyes Univ 10. https://doi.org/citeulike-article-id:6592152
15. Geem ZW (2001) A new heuristic optimization algorithm: harmony search. Simulation
16. Yang XS (2012) Flower pollination algorithm for global optimization. Lect Notes Comput Sci (including Subser Lect Notes Artif Intell Lect Notes Bioinformatics) 7445:240–249. https://doi.org/10.1007/978-3-642-32894-7_27
17. Rashedi E, Nezamabadi-pour H, Saryazdi S (2009) GSA: a gravitational search algorithm. Inf Sci (NY) 179:2232–2248. https://doi.org/10.1016/j.ins.2009.03.004
18. Tan KC, Chiam SC, Mamun AA, Goh CK (2009) Balancing exploration and exploitation with adaptive variation for evolutionary multi-objective optimization. Eur J Oper Res 197:701–713. https://doi.org/10.1016/j.ejor.2008.07.025
19. Alba E, Dorronsoro B (2005) The exploration/exploitation tradeoff in dynamic cellular genetic algorithms. IEEE Trans Evol Comput 9:126–142. https://doi.org/10.1109/TEVC.2005.843751
20. Paenke I, Jin Y, Branke J (2009) Balancing population- and individual-level adaptation in changing environments. Adapt Behav 17:153–174. https://doi.org/10.1177/105971230910 3566
21. Cuevas E, Echavarría A, Ramírez-Ortegón MA (2014) An optimization algorithm inspired by the states of matter that improves the balance between exploration and exploitation. Appl Intell 40:256–272. https://doi.org/10.1007/s10489-013-0458-0
22. Shilane D, Martikainen J, Dudoit S, Ovaska SJ (2008) A general framework for statistical performance comparison of evolutionary computation algorithms. Inf Sci (NY) 178:2870–2879. https://doi.org/10.1016/j.ins.2008.03.007
23. Elbeltagi E, Hegazy T, Grierson D (2005) Comparison among five evolutionary-based optimization algorithms. Adv Eng Inform 19:43–53. https://doi.org/10.1016/j.aei.2005.01.004
24. García S, Molina D, Lozano M, Herrera F (2009) A study on the use of non-parametric tests for analyzing the evolutionary algorithms' behaviour: a case study on the CEC'2005 special session on real parameter optimization. J Heuristics 15:617–644. https://doi.org/10.1007/s10 732-008-9080-4
25. Shlesinger MF, Zaslavsky GM, Frisch U (1995) Lévy flights and related topics in physics. Springer, Berlin Heidelberg, Berlin, Heidelberg
26. Barthelemy P, Bertolotti J, Wiersma DS (2008) A Lévy flight for light. Nature 453:495–498. https://doi.org/10.1038/nature06948
27. Das S, Konar A (2007) A swarm intelligence approach to the synthesis of two-dimensional IIR filters. Eng Appl Artif Intell 20:1086–1096. https://doi.org/10.1016/j.engappai.2007.02.004
28. Mladenov VM, Mastorakis NE (2001) Design of two-dimensional recursive filters by using neural networks. IEEE Trans Neural Netw 12:585–590. https://doi.org/10.1109/72.925560
29. Mastorakis NE, Gonos IF, Swamy MNS (2003) Design of two-dimensional recursive filters using genetic algorithms. Circuits Syst I Fundam Theory Appl IEEE Trans 50:634–639. https://doi.org/10.1109/TCSI.2003.811019
30. Tsai J-T, Ho W-H, Chou J-H (2009) Design of two-dimensional IIR digital structure-specified filters by using an improved genetic algorithm. Expert Syst Appl 36:6928–6934. https://doi.org/10.1016/j.eswa.2008.08.065
31. Sarangi SK, Panda R, Dash M (2014) Design of 1-D and 2-D recursive filters using crossover bacterial foraging and Cuckoo search techniques. Eng Appl Artif Intell 34:109–121. https://doi.org/10.1016/j.engappai.2014.05.010
32. Cuevas E, Cienfuegos M (2014) A new algorithm inspired in the behavior of the social-spider for constrained optimization. Expert Syst Appl 41:412–425. https://doi.org/10.1016/j.eswa.2013.07.067

33. Lin Y-L, Chang W-D, Hsieh J-G (2008) A particle swarm optimization approach to nonlinear rational filter modeling. Expert Syst Appl 34:1194–1199. https://doi.org/10.1016/j.eswa.2006.12.004
34. Cuevas E, Gálvez J, Hinojosa S et al (2014) A comparison of evolutionary computation techniques for IIR model identification. J Appl Math 2014. https://doi.org/10.1155/2014/827206
35. Hochberg Y (1988) A sharper Bonferroni procedure for multiple tests of significance. Biometrika 75:800–802. https://doi.org/10.1093/biomet/75.4.800
36. Cuevas E, González A, Fausto F, Zaldívar D, Pérez-Cisneros M (2015) Multithreshold segmentation by using an algorithm based on the behavior of locust swarms. Math Probl Eng 2015:805357
37. Cuevas E, Zaldivar D, Pérez-Cisneros M (2011) Seeking multi-thresholds for image segmentation with learning automata. Mach Vis Appl 22(5):805–818
38. Ramírez-Ortegón MA, Tapia E, Ramírez-Ramírez LL, Rojas R, Cuevas E (2010) Transition pixel: a concept for binarization based on edge detection and gray-intensity histograms. Pattern Recogn 43(4):1233–1243

Chapter 3
Comparison of Metaheuristics for Chaotic Systems Estimation

In recent years Parameter Estimation (PE) has attracted the attention of the scientific community. PE can be applied in diverse areas of engineering and science. An important field of study is system identification of Chaotic Systems (CS) to synchronize and control the chaos. The PE of CS is a highly non-linear optimization problem within a multi-dimensional space where classic optimization techniques are not suitable to use. Evolutionary Computation Techniques (ECT) are commonly used to solve complex non-linear optimization problems. Recently, some traditional and modern ECT have been introduced to estimate the parameters for chaotic systems nonetheless, the reported results and conclusions in the published literature are based only on the fitness function of each ECT despite their solutions. This chapter presents a comparative examination of the most used ECTs commonly used to solve the PE problem of chaotic systems not only based on the performance criteria but also on their solution homogeneity. Results over the Lorenz and Chen chaotic systems are investigated and statistically validated by employing non-parametric tests.

3.1 Introduction

Parameter estimation (PE) is an essential task in many disciplines of engineering. However, most of the parameters to be estimated are associated with non-linear and complex applications. These applications include communication, control systems, and signal processing [1–3]. Recently PE has been applied broadly to identify the best arrangement of a Chaotic System (CS). In adaptive control, chaos is defined as an unstable dynamic behavior that exhibits sensitive dependence on initial conditions. These kinds of systems are regulated and synchronized employing techniques of feedback control. In the related literature there exist two well-known CS: Lorenz [4] and Chen [5], that are commonly used to control trajectories. However, the principal difficulty is obtaining the best parameter estimation to control the system. Such a problem can be formulated as a multidimensional optimization problem.

© The Author(s), under exclusive license to Springer Nature Switzerland AG 2023   37
E. Cuevas et al., *Analysis and Comparison of Metaheuristics*, Studies in Computational
Intelligence 1063, https://doi.org/10.1007/978-3-031-20105-9_3

The PE task consists in adopting a strategy to automatically discover the parameters of an adaptive model to obtain an optimal model for an unknown chaotic system. Considering the PE problem as an optimization problem it is required to minimize an error function among the output of the estimated model and the output of the real model. In other words, the reduction of the error delimits the quality of the estimated set of parameters. The quality is directly reflected in the optimal model [6]. Classically, gradient-based methods as the least mean square and its modifications have been widely used to resolve this problem [7]. However, the least mean square techniques are only suitable for the model structure of systems having the feature of being linear in the parameters. Due to the inherent nature of CS, the nonlinearity causes the optimization process more complex concerns the size of parameter space [8], the presence of local minima and the sensitivity of the objective function to each parameter model. The associated complications with the application of gradient-based optimization methods for resolving this class of engineering problems have provided to the development of alternative resolutions. In this context, ECTs have received the attention of researchers concerning their potential as global optimization methods in real-world applications. ECT is a collection of algorithms that are based on collective intelligence. In different words, ECTs use a population of individuals in which the members, exchange information to achieve the global optima. These methods are commonly motivated in natural processes. Some ECT are the Particle Swarm Optimization (PSO) [9], Artificial Bee Colony (ABC) [10], Harmony Search (HS) [11], Differential Evolution (DE) [12], Cuckoo Search Algorithm (CSA) [13] and Gravitational Search Algorithm (GSA) [14]. As an alternative to traditional gradient-based techniques, the use of ECTs to estimate the parameters for chaotic systems has generated a robust CS identification. Such methods use different algorithms like PSO [8] and DE [15] just to mention few of them.

The application of ECTs depends on the conditions of particular problems because no single optimization algorithm can solve each problem competitively. Consequently, when novel alternative approaches are introduced, their performance must be suitably evaluated. Commonly, comparisons are done applying synthetic numerical benchmark problems, they help to verify if a particular algorithm overlooks any statistical test. However in the related literature there exist a few group of comparative studies of various evolutionary algorithms, for example the proposed in [16]. Therefore, it is important to examine and analyze the performance of ECTs from an application point of view.

In this chapter, it is presented a comparison of several ECTs that are applied to estimate chaotic systems parameters. In the comparison, special attention is paid to the recently proposed methods: Gravitational Search Algorithm (GSA) and Cuckoo Search Algorithm (CSA). Moreover, well-known methods are included such as Particle Swarm Optimization (PSO), Artificial Bee Colony (ABC), Harmony Search (HS) and Differential Evolution (DE). Outcomes over two distinct CS are presented and validated within a statistically significant framework.

The remains of this chapter are prepared as follows: In Sect. 3.2, ECTs are introduced. Section 3.3 gives a concise description of parameter estimation for chaotic systems. In Sect. 3.4 the experimental results and the statistical validation of the solutions for each algorithm proposed are performed. Finally, some conclusions are presented in Sect. 3.5.

3.2 Evolutionary Computation Techniques (ECT)

Real-world optimization problems are generally considered black-box modeling. When this type of situation is examined, less information is stored and the low knowledge of the system presents the parameter estimation hard to solve. In the worst scenario, there is no present single characteristic of the fitness function. One of the advantages of ECT is to identify the optimum solution. They can simply adapt their operators to black-box formulation and extremely difficult search spaces. Most of the ECTs are population-based methods, which indicates that a particular set of candidate solutions exchange information through the optimization procedure. This population is initialized in a random fashion and at each iteration, the solutions evolve towards more suitable search regions. This mechanism is independently operated by the own ECT. ECT has been also employed to resolve many optimization problems in various disciplines of science, this fact verifies that ECTs are powerful alternatives to solve complex or black-box modeling. ECTs deals with several optimization problems such as: multimodal optimization, dynamic optimization and multi-objective optimization. [17–19].

3.2.1 Particle Swarm Optimization (PSO)

The PSO method, developed by Kennedy and Eberhart in 1995 [9], is a population-based stochastic search technique that is inspired on the social behavior of fish schooling or bird flocking. This approach finds the optimal value considering a set or swarm formed by candidate solutions of the problem to be solved, which are known as particles. From an implementation perspective, in the PSO procedure, a population $\mathbf{p}^k(\{\mathbf{p}_1^k, \mathbf{p}_2^k, ..., \mathbf{p}_N^k\})$ of N individuals iteratively are processed from the initial point ($k = 0$) to a maximum gen number of iterations ($k = gen$). Each individual $\mathbf{p}_i^k$ ($i \in [1, ..., N]$) corresponds to a d-dimensional structure $\{\mathbf{p}_{i,1}^k, \mathbf{p}_{i,2}^k, ..., \mathbf{p}_{i,d}^k\}$ where each decision variable corresponds to a dimension of the optimization problem to be solved. The fitness value of each individual $\mathbf{p}_i^k$ is obtained by the evaluation of the cost function $f(\mathbf{p}_i^k)$ whose final value corresponds to the fitness value of $\mathbf{p}_i^k$, through the evolution process, the best global location $\mathbf{g}(g_1, g_2, ..., g_d)$ been so far is stored with the best position $\mathbf{p}_i^*(\mathbf{p}_{i,1}^*, \mathbf{p}_{i,2}^*, ..., \mathbf{p}_{i,d}^*)$.

In this chapter, the varied PSO version developed by Lin in [20] has been employed. Below such proposal, the new position $\mathbf{p}_i^{k+1}$ of each individual $\mathbf{p}_i^k$ is computed by the following mathematical expressions:

$$v_{i,j}^{k+1} = \omega \cdot v_{i,j}^k + c_1 \cdot r_1 \cdot (p_{i,j}^* - p_{i,j}^k) + c_2 \cdot r_2 \cdot (g_j - p_{i,j}^k),$$
$$p_{i,j}^{k+1} = p_{i,j}^k + v_{i,j}^{k+1} \tag{3.1}$$

where ω is known as the inertia weight which controls the velocity ($i \in [1, ..., N]$, $j \in [1, ..., d]$). c_1 and c_2 are the acceleration coefficients that control the behavior of each particle towards the positions $\mathbf{g}$ and $\mathbf{p}_i^*$, respectively. r_1 and r_2 random numbers over the range: $[0, 1]$.

3.2.2 Artificial Bee Colony (ABC)

The artificial bee colony (ABC) algorithm, developed by Karaboga [10], is an ECT based on the clever foraging action of honeybee swarms. In the ABC operation, a population $\mathbf{L}^k(\{\mathbf{l}_1^k, \mathbf{l}_2^k, ..., \mathbf{l}_N^k\})$ of N food locations (solutions) is modified considering the initial point ($k = 0$) to a maximum number of iterations gen ($k = gen$). Each food location $\mathbf{l}_i^k$ ($i \in [1, ..., N]$) corresponds to a d-dimensional structure $\{\mathbf{l}_{i,1}^k, \mathbf{l}_{i,2}^k, ..., \mathbf{l}_{i,d}^k\}$ where each decision variable represents to a variable of the optimization problem to be solved. Subsequent the initialization process, a cost function computes each food position to assess whether it represents a satisfactory solution (nectar-amount) or not. Conducted by the values of such a cost function, the candidate solution $\mathbf{l}_i^k$ is evolved based on different honeybee operations. For the principal evolutionary operator, each food position $\mathbf{l}_i^k$ obtains a new food source $\mathbf{t}_i$ in the neighborhood of its current position:

$$\mathbf{t}_i = \mathbf{l}_i^k + \phi(\mathbf{l}_i^k - \mathbf{l}_r^k), \quad i, r \in (1, 2, ..., N), \tag{3.2}$$

where $\mathbf{l}_r^k$ is a random food position, which satisfies the condition ($r \neq i$). The scale factor ϕ is a random value among the range $[-1, 1]$. Once a new solution $\mathbf{t}_i$ is obtained, a cost value will correspond to the associated probability with a solution $\mathrm{fit}(\mathbf{l}_i^k)$. The cost value for a minimization process can be indicated to a candidate location $\mathbf{l}_i^k$ by the subsequent expression:

$$\mathrm{fit}(\mathbf{l}_i^k) = \begin{cases} \frac{1}{1 + f(\mathbf{l}_i^k)}, & \text{if } f(\mathbf{l}_i^k) \geq 0, \\ 1 + \mathrm{abs}(f(\mathbf{l}_i^k)), & \text{if } f(\mathbf{l}_i^k) < 0, \end{cases} \tag{3.3}$$

where $f(\cdot)$ corresponds to the cost value of the cost function to be minimized. Once the fitness values are computed, a greedy selection mechanism is employed among $\mathbf{t}_i$ and $\mathbf{l}_i^k$. If $\mathrm{fit}(\mathbf{t}_i)$ is better than $\mathrm{fit}(\mathbf{l}_i^k)$ then the candidate solution $\mathbf{l}_i^k$ is replaced by $\mathbf{t}_i$; contrarily, $\mathbf{l}_i^k$ persists.

3.2.3 Cuckoo Search (CS)

CS is one of the modern nature-inspired algorithms that has been implemented by Yang and Deb [21]. CS is based on the brood parasitism of some cuckoo species. In addition, this algorithm enhances its performance by the so-called Lévy flights [22], rather than by simple isotropic random walks. From the implementation perspective of the CS procedure, a population $\mathbf{E}^k(\{\mathbf{e}_1^k, \mathbf{e}_2^k, ..., \mathbf{e}_N^k\})$ of N solutions (eggs) is evolved from the starting point $(k = 0)$ to a maximum *gen* number of iterations $(k = 2 \cdot gen)$. Each solution $\mathbf{e}_i^k (i \in [1, ..., N])$ corresponds to a d-dimensional vector $\{e_{i,1}^k, e_{i,2}^k, ..., e_{i,d}^k\}$ where each variable corresponds to a parameter to be found in the optimization process. The quality of each solution $\mathbf{e}_i^k$ is evaluated by the cost function $f(\mathbf{e}_i^k)$ whose final result represents the fitness value of $\mathbf{e}_i^k$. Under this approach, three separate operators represents the evolution process of CS: (a) *Lévy flight*, (b) *the replacing of nests operator for constructing new solutions*, and (c) *the elitist selection strategy*.

(a) *The Lévy Flight*. One of the most crucial features of the cuckoo search is the application of Lévy flights to produce newer potential solutions. In this mechanism, a new candidate solution $\mathbf{e}_i^{k+1}$ is obtained by the perturbation mechanism over the current $\mathbf{e}_i^k$ with a change of location c_i. To compute c_i, a random value s_i a variable mechanism is generated by following a symmetric Lévy distribution. For producing s_i, Mantegna's algorithm [23] is applied as follows:

$$s_i = \frac{\mathbf{u}}{|\mathbf{v}|^{1/\beta}} \tag{3.4}$$

where $\mathbf{u}(\{u_1, ..., u_d\})$ and $\mathbf{v}(\{v_1, ..., v_d\})$ are n-dimensional structures and $\beta = 3/2$. Each value of $\mathbf{u}$ and $\mathbf{v}$ is computed following the normal distribution function:

$$u \sim N(0, \sigma_u^2), \quad v \sim N(0, \sigma_v^2),$$
$$\sigma_u = \left(\frac{\Gamma(1+B) \cdot \sin(\pi \cdot \beta/2)}{\Gamma((1+\beta)/2) \cdot \beta \cdot 2^{(\beta-1)/2}} \right)^{1/\beta}, \quad \sigma_v = 1 \tag{3.5}$$

where $\Gamma(\cdot)$ corresponds to the Gamma distribution function. Once s_i has been computed, c_i is calculated according the following expression:

$$c_i = 0.01 \cdot s_i \oplus (\mathbf{e}_i^k - \mathbf{e}^{best}) \tag{3.6}$$

where $\oplus$ denotes entry wise multiplication operations whereas $\mathbf{e}^{best}$ is the best solution been so far during the optimization procedure. Lastly, the new solution $\mathbf{e}_i^{k+1}$ is computed based on the following structure:

$$\mathbf{e}_i^{k+1} = \mathbf{e}_i^k + c_i \tag{3.7}$$

(b) *Replacing Some Nests by Constructing New Solutions.* Below this process, a collection of individuals is selected and restored with a new cost value. Each solution e_i^k can be selected considering a probability value $p_a \in [0, 1]$. To implement this mechanism, the following rules are applied: if a random number r_1 is less than p_a, the solution e_i^k is then chosen and altered by the mathematical expression from Eq. 3.4; contrarily e_i^k continues with no change. This process can be summarized by the following mathematical structure:

$$e_i^{k+1} = \begin{cases} e_i^k + \text{rand} \cdot \left(e_j^k - e_h^k \right) & \text{with probability } p_a \\ \quad\quad e_i^k & \text{with probability } (1 - p_a) \end{cases} \quad (3.8)$$

where rand is a random number. j and h are random integers from 1 to N.

(c) *The Elitist Selection Strategy.* Later, obtaining e_i^{k+1} either by the operator (a) or by the operator (b), it requires to be compared through its past value e_i^k. If the fitness value of e_i^{k+1} is better than e_i^k, then e_i^{k+1} is accepted as the final solutions; otherwise, e_i^k is stored. This approach can be represented by the following equation:

$$e_i^{k+1} = \begin{cases} e_i^{k+1} & \text{if } f\left(e_i^{k+1}\right) < f\left(e_i^k\right) \\ \quad e_i^k & \text{otherwise} \end{cases} \quad (3.9)$$

The elitist mechanism rule suggests that only high-quality solutions which are the most related to the host bird´s solutions have the chance to survive and became adult cuckoos.

3.2.4 Harmony Search (HS)

The HS is an optimization search approach based on music improvisation presented by Geem in [11]. It is motivated by the perception of music for a precise environment of harmony using the improvisation as a metaphor. The metaphor within the optimization process consists by the process to search the optimal value of a certain optimization problem. When musicians improvise, they develop three commonly used rules to get a genuine melody: plays any important piece of music accurately from his memory, plays something comparable to a recognized piece and produce new or plays arbitrary notes. In the HS evolutionary mechanism those rules are idealized as: (a) *Harmony Memory (HM)*, (b) *pitch adjustment* and (c) *initialize a random solution.*

(a) *Harmony Memory (HM) Consideration.* This phase will assure that the best compositions will be conducted over new harmony memory. In HM method, a parameter to help the selection mechanism of the elements of the composition is required. This configuration parameter is called Harmony Memory Consideration Rate *(HMCR)* and is defined as $HMCR \in [0, 1]$. If this parameter value is too low, only some best pieces are selected and it may converge too slowly. Contrary, if this parameter value is extremely high, then all the harmonies will be used in HM composition, then other harmonies will not be considered in the exploration process. In resume, the new harmony could be an element of the HM or randomly generated, depending on the *HMCR* parameter value.

(b) *Pitch Adjustment.* Each component previously obtained by the memory consideration rule is further analyzed to determine if it should be pitch-adjusted or not. For this process, the parameter called Pitch-Adjusting Rate *(PAR)* is defined to assign the frequency of the adjustment and the Bandwidth factor *(BW)* to control the local search around the neighborhood of the selected elements of the HM. Therefore, the pitch adjusting decision rule is mathematically composed by:

$$x_{new} = \begin{cases} x_{new} = x_{new} \pm \text{rand}(0, 1) \cdot BW & \text{with probability } PAR \\ x_{new} & \text{with probability } (1 - PAR) \end{cases}$$

$$(3.10)$$

Pitch adjusting is capable of producing new possible harmonies by lightly adjusting original variable locations. Such operation can be viewed as similar to the mutation process in genetic algorithms. Hence, the decision variable is either perturbed in terms of a random value among the range 0 and *BW* or not. To protect the pitch adjusting operation, it is important to guarantee that points prevailing outside the possible range must be relocated.

(c) *Initialize a random solution.* This operation seeks to improve the diversity of the solutions in the optimization problem. As is mentioned in (a) if a new solution is not chosen from the harmony memory, it must be generated based on a random process within the entire search space. This mechanism helps to explore new wider search zones avoiding the local optimal. This process is summarized according to Eq. 3.11.

$$x_{new} = \begin{cases} x_i \in \{x_1, x_2, \dots, x_{HMS}\} & \text{with probability } HMCR \\ l + (u - l \cdot \text{rand}(0, 1) & \text{with probability } 1 - HMCR \end{cases}$$

$$(3.11)$$

where *HMS* corresponds to the size of the HM, x_i is considered as an element chosen from the HM. And l and u represents the lower and upper limits in the search space, respectively. Lastly, rand is a random value following a uniform distribution among 0 and 1.

3.2.5 *Differential Evolution (DE)*

The DE is an optimization vector-based search algorithm implemented by Storn and Price [24]. Under the DE structure, a population of individuals (vectors) is created containing the vectors (feasible solutions), each solution can be defined as follows:

$$\mathbf{x}_i^t = (x_{1,i}^t, x_{2,i}^t, ..., x_{d,i}^t), \; i = 1, 2, ..., n \tag{3.12}$$

where $\mathbf{x}_i^t$ corresponds to the i-th vector in current iteration t. Then, n represents the d-dimensional potential solutions. The original structure of the DE algorithm considers three main phases: (*a*) *mutation*, (*b*) *crossover* and (*c*) *selection*.

(a) *Mutation.* This procedure is employed over a vector solution $\mathbf{x}_i^t$ in current t generation. At first, there are selected three different vectors by a random fashion from the population of individuals: $\mathbf{x}_p$, $\mathbf{x}_q$ and $\mathbf{x}_r$, then considering the mutation procedure exhibit in Eq. 3.13 generate a fourth vector.

$$\mathbf{v}_i^{t+1} = \mathbf{x}_p^t + F(\mathbf{x}_q^t - \mathbf{x}_r^t) \tag{3.13}$$

where $F \in [0, 1]$ corresponds to a parameter commonly referred as the differential weight and this parameter is adjusted considering the problem to be solved.

(b) *Crossover.* The application of this evolutionary operator over the population of the DE method, is conducted by the crossover rate parameter $C_r \in [0, 1]$. The C_r helps to adjust few candidate vectors of the population and it is employed at each dimension of a solution. The crossover starts using the generation of random individuals $r_i \in [0, 1]$ and then the j-th variable of $\mathbf{v}_i$ is modified as follows:

$$\mathbf{u}_{j,i}^{t+1} = \begin{cases} \mathbf{v}_{j,i} & \text{if} \;\; r_i \leq C_r \\ \mathbf{x}_{j,i}^t & \text{otherwise} \end{cases}, \; j = 1, 2, ..., d, \; i = 1, 2, ..., n \tag{3.14}$$

(c) *Selection.* The selection strategy is basically the exact procedure that is employed in Genetic Algorithms (GA) [25]. Considering the cost value in a minimization scenario, the solution that contains the lowest fitness is chosen (Eq. 3.15) and for a maximization scenario is a requirement to adjust the sign to get the solution with the higher fitness.

$$\mathbf{x}_i^{t+1} = \begin{cases} \mathbf{u}_i^{t+1} & \text{if} \; f(\mathbf{u}_i^{t+1}) \leq f(\mathbf{x}_i^t) \\ \mathbf{x}_i^t & \text{otherwise} \end{cases} \tag{3.15}$$

3.2.6 Gravitational Search Algorithm (GSA)

GSA is a recent developed metaheuristic optimization method which is based by the Newtonian´s law of gravity and motion. It was presented in 2009 by Rashedi [14] and it is composed by a set of search agents known as masses. All the masses from the population, attract among each other caused by the gravity force. This force conducts a movement interaction among all masses towards the objects containing heavier masses. Considering an algorithmic description for GSA the i-th individual is constructed as:

$$\mathbf{x}_i = (x_i^1, ..., x_i^d, ..., x_i^n) \text{ for } i = 1, 2, ..., N \tag{3.16}$$

where x_i^d corresponds to the position of i-th solution in the d-dimension. At a current time t, the acting force on mass i from mass j is computed as follows:

$$F_{ij}^d(t) = G(t)\frac{Mp_i(t) \times Ma_j(t)}{R_{ij}(t) + \varepsilon}(x_j^d(t) - x_i^d(t)) \tag{3.17}$$

where Ma_j corresponds the active gravitational mass related to agent j, Mp_i represents the passive gravitational of agent i, $G(t)$ corresponds to the gravitational constant at time t, ε is a small constant and $R_{ij}(t)$ is the Euclidian distance between i-th and j-th individuals. The total force that affects the entire system over a given solution i within a d dimensional search space is calculated as:

$$F_i^d(t) = \sum_{j=1, j\neq i}^{N} \text{rand}_j F_{ij}^d(t) \tag{3.18}$$

Hence, by the Newton's second law the acceleration of the agent i at time t is given as:

$$a_i^d(t) = \frac{F_i^d(t)}{Mn_i(t)} \tag{3.19}$$

where $Mn_i(t)$ represents the inertial mass of the individual i. Therefore, the next velocity and position of a new candidate solution can be defined as:

$$\begin{aligned} v_i^d(t+1) &= \text{rand}_i \times v_i^d(t) + a_i^d(t) \\ x_i^d(t+1) &= x_i^d(t) + v_i^d(t+1) \end{aligned} \tag{3.20}$$

3.3 Parameter Estimation for Chaotic Systems (CS)

Chaotic systems are extremely nonlinear, under such reason many nonlinear control techniques can be applied to synchronize those chaotic systems such as the ones reported in Hegazi [26] and Huang [27]. Nevertheless, these methods are invalid when the parameters of CS are unknown. This is the principal cause why parameter estimation (PE) for chaotic systems has brought the care of many researchers all over the world. To determine the parameters of a CS, it is important to consider the following n- dimensional nonlinear model:

$$\dot{X} = F(\mathbf{X}, \mathbf{X}_0, \boldsymbol{\theta}) \tag{3.21}$$

where $\mathbf{X} = [x_1, x_2, ..., x_n]^T$ represents the position vector, $\mathbf{X}_0$ indicates the initial state vector and $\boldsymbol{\theta} = [\theta_1, \theta_2, ..., \theta_n]^T$ is the initial vector of parameters and F is an assigned nonlinear vector function. To determine the values of the parameters in Eq. 3.21, an expected model is described as:

$$\dot{\hat{X}} = F\left(\hat{\mathbf{X}}, \mathbf{X}_0, \hat{\theta}\right) \tag{3.22}$$

where $\hat{\mathbf{X}} = [\hat{x}_1, \hat{x}_2, ..., \hat{x}_n]^T$ represents the approximated parameter vector and $\hat{\theta} = [\hat{\theta}_1, \hat{\theta}_2, ..., \hat{\theta}_n]^T$ the estimated parameters of the estimated nonlinear system $\dot{\hat{X}}$. As the evolutionary algorithms depends on the cost function value to guide the searching method towards the optimal solution, it must be accurately defined. In this chapter, the Mean Square Error (MSE) is considered as the objective function. It lists the performance for each possible solution. The MSE examines the real and the predicted responses for a given number of units M. The MSE is defined as follows:

$$J(\theta) = \frac{1}{M} \sum_{k=1}^{M} \left(X(k) - \hat{X}(k)\right)^2 \tag{3.23}$$

where $X(k)$ represents the original values and $\hat{X}(k)$ corresponds to the estimated values at current time k. The main objective of this, consist of minimize the cost function $J(\theta)$ by adjusting θ. The optimal value $\theta*$ is obtained when the error function $J(\theta)$ reaches the minimum value, this procedure is mathematically described as follows:

$$\theta^* = \arg \min(J(\theta)) \tag{3.24}$$

The parameter estimation problem for chaotic systems can be considered as an optimization problem which is described in Fig. 3.1.

Fig. 3.1 Generic procedure of a metaheuristic approach

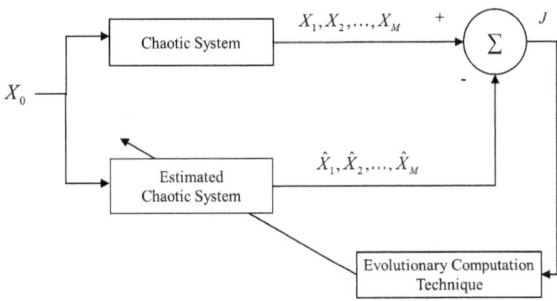

3.4 Experimental Results

In the comparative analysis introduced in this chapter, a collection of experiments has been proposed to evaluate the performance of each evolutionary approach. For the tests, two different chaotic systems are analyzed; the Chen system and the Lorenz system. The mathematical representation of the Lorenz system is defined as:

$$
\begin{aligned}
\frac{dx}{dt} &= \sigma(y - x) \\
\frac{dy}{dt} &= rx - y - xz \\
\frac{dz}{dt} &= xy - bz
\end{aligned}
\tag{3.25}
$$

where $\sigma = 10$, $r = 28$ and $b = 8/3$ represents the original parameter values. On the other hand, the Chen system is mathematically described as:

$$
\begin{aligned}
\frac{dx}{dt} &= a(y - x) \\
\frac{dy}{dt} &= (b - a)x + by - xz \\
\frac{dz}{dx} &= -cz + xy
\end{aligned}
\tag{3.26}
$$

where $a = 35$, $b = 28$ and $c = 3$ corresponds to the original parameter values. Both initial conditions for each system were chosen as: [1]. The amount of states for both the original and the approximated system is configured to 100.

The initial parameter configuration for each metaheuristic method considered in the comparison is described as follows:

(1) PSO: the evolutionary parameters are configured to $c_1 = 2$, $c_2 = 2$; besides, the weight factor decreases linearly from 0.9 to 0.2 [8].
(2) ABC: the ABC method has been coded using the reported guidelines according to [10].

(3) DE: the implemented version was DE/rand/bin where $cr = 0.5$ and *differential weight* $= 0.2$ [24].
(4) HS: the evolutionary method was implemented considering the reported guidelines from [11] with $HMCR = 0.95$ and $PAR = 0.3$.
(5) CSA: corresponding to [21] and [28], the configuration parameters are set to $p_a = 0.25$ and the number of generations $gen = 100$.
(6) GSA: the evolutionary approach has been coded from [14].

The size of the population for each of the six algorithms has been configured as 25 whereas the maximum number of generations has been set to 100. The experimental results are separated into two different sections. The first section shows the performance of elected algorithms to estimate the parameter of the Lorenz and Chen systems individually (one dimension) and in groups (two or three dimensions). Meantime, in the second section, the experimental results are investigated from a statistical perspective by the Wilcoxon Rank Test and validating those experimental results by the Bonferroni correction test.

3.4.1 Chaotic System Parameter Estimation

The reported results consider six different experiments for each evolutionary method. Such experiments include the parameter identification of one-dimensional, two-dimensional and three-dimensional parameter estimation using the Chen and Lorenz chaotic systems.

3.4.1.1 One-Dimensional Parameter Estimation

At first, it is considered the analysis for only one parameter of the Lorenz system to be approximated. The experimental results for this experiment are presented in Table 3.1. Also, it is considered that just one parameter of the Chen system must be approximated. The experimental results for this study are presented in Table 3.2.

Table 3.1 Performance results of one-dimensional problem for Lorenz system

	PSO	ABC	DE	HS	CSA	GSA
σ	10.3415	10.0002	9.9708	10.0082	10.0000	10.0000
J	2.0373e−10	4.5666e−12	2.6443e−07	1.2465e−08	2.6767e−19	**4.2689e−23**
r	28.7378	27.9946	27.9520	28.0135	28.0000	28.0000
J	2.0972e−07	3.2128e−08	2.0873e−06	3.0226e−07	1.3071e−19	**7.7107e−22**
b	3.0000	2.6636	2.6551	2.6705	2.6667	2.6667
J	2.0277e−08	1.6602e−09	4.4504e−08	1.7030e−08	7.0267e−20	**1.3411e-22**

Table 3.2 Performance results of one-dimensional problem for Chen system

	PSO	ABC	DE	HS	CS	GSA
a	35.3506	35.0053	34.9924	34.9935	35.0000	35.0000
J	3.6998e−09	3.7128e−09	2.0310e−05	4.1511e−08	2.9510e−19	**4.1632e−22**
b	28.3041	28.0000	28.0129	27.9989	28.0000	28.0000
J	2.7571e−06	1.6724e−08	9.9146e−06	1.1411e−08	1.9136e−17	**3.0693e−21**
c	3.1174	2.9990	2.9998	2.9951	3.0000	3.0000
J	1.0765e−09	2.7945e−10	1.9080e−08	6.7298e−10	1.0451e−19	**1.9135e−23**

Table 3.1 presents the numerical results taken from the evolutionary methods: PSO, ABC, DE, HS, CSA and GSA. These approaches are tested considering the Lorenz system. The test is performed by modifying each parameter of the CS from the other parameters while the rest of them are set to a predefined value. The purpose of this test is to support the performance of the algorithms in one dimension. It is essential to consider that the number of generations for this experiment was established to 100. According to the numerical results, the GSA is the process which performs better in the estimation of the Lorenz System considering one dimension.

Correspondingly to Tables 3.1 and 3.2 illustrates the performance results of the chosen evolutionary approaches to estimate each parameter of the Chen chaotic system. Regarding the experimental results of the cost function is possible to notice that GSA algorithm obtains the minimum value for this problem. To investigate the evolution of the cost function values during the optimization process, in Fig. 3.2a the values for the Lorenz system are graphically displayed. Meanwhile, in Fig. 3.2b the cost function values acquired for the Chen system are plotted. From these figures is viable to state that the GSA converges in fewer generations than the remaining evolutionary algorithms.

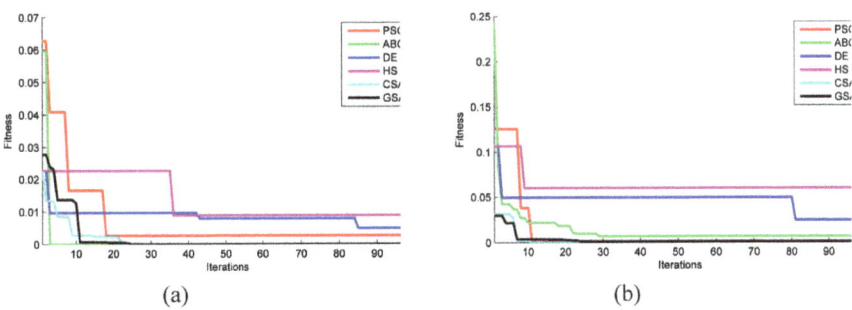

(a)   (b)

Fig. 3.2 Evolution process of the cost function value for a one-dimensional problem obtained by PSO, ABC, DE, HS, CSA and GSA for **a** Lorenz CS and **b** Chen CS

3.4.1.2 Two-Dimensional Parameter Estimation

In this experimentation stage both the Lorenz and Chen chaotic systems, are considered to find pairs of parameters. Considering this, the Lorenz systems consider the combination of its parameters according to the Eq. 3.27, then each evolutionary method operates with a two-dimensional optimization problem that increases the complexity of the search space. The experimental results for the implementation of the specified methods for this experimentation regarding the Lorenz chaotic system are displayed in Table 3.3.

$$[\sigma, r], [\sigma, b] \text{ and } [b, r] \qquad (3.27)$$

Table 3.3 shows the experimental results acquired by the selected algorithms involved to estimate the parameters of the Lorenz CS, the experiment was performed regarding three different couples of parameters. This experimentation is conducted to examine the efficiency of the compared ECT in a two-dimensional parameter estimation problem. The composition of each couple of elements to be calculated is described in Eq. 3.28. For this experiment, the amount of generations was set to 100 for all the evolutionary techniques.

$$[a, b], [b, c] \text{ and } [a, c] \qquad (3.28)$$

Table 3.4 displays the experimental results acquired by the experimentation of the PSO, ABC, DE, HS, CSA and GSA methods to estimate couples of parameters from the Chen CS. The investigation employing the Chen CS was conducted considering the same care as the Lorenz CS. From the outcomes of Table 3.3, it can be noticed that the outcomes of the GSA algorithm for the Lorenz system are better than the rest of the evolutionary techniques. Meanwhile, in Table 3.4, using the CSA for parameters a and b of the Chen CS are the best. However, for the remainder of the combination of parameters, the GSA holds its efficiency. The evolution of the

Table 3.3 Performance results of two-dimensional problem for Lorenz system

	PSO	ABC	DE	HS	CSA	GSA
σ	10.2293	10.0397	9.9343	10.1521	10.0000	10.0000
r	28.3408	28.0341	28.0136	27.9398	28.0000	28.0000
J	1.4585e−04	9.4979e−05	1.1927e−04	2.4674e−05	7.3093e−12	**2.6085e−21**
σ	10.6089	9.8557	10.0307	9.9680	10.0000	10.0178
b	3.0000	2.7073	2.6575	2.6792	2.6667	2.6609
J	6.6248e−07	1.7635e−05	1.2946e−05	2.0935e−05	1.5621e−12	**1.2385e−19**
b	3.0000	2.4929	2.6851	2.6434	2.6666	2.6667
r	28.3513	28.1535	27.8856	28.0001	28.0001	28.0000
J	0.0011	6.3016e−05	9.5844e−05	1.5472e−04	3.7434e−12	**9.3895e−21**

Table 3.4 Performance results of two-dimensional problem for Chen system

	PSO	ABC	DE	HS	CSA	GSA
a	35.9623	36.0448	34.8281	36.3564	35.0003	35.2291
b	29.4769	28.5689	27.9031	28.6719	28.9992	28.1116
J	5.6641e−04	6.0294e−04	2.3821e−04	3.0392e−04	**6.0028e−12**	1.2149e−06
b	28.0709	28.0301	28.1287	28.0202	28.0000	27.9983
c	3.8663	3.0461	2.7461	2.9995	3.0000	3.0024
J	4.5480e−04	8.6206e−05	1.1307e−04	1.9767e−04	3.7080e−11	**1.6683e−19**
a	35.3818	34.9870	34.8585	34.8587	35.0003	35.0000
c	3.7622	3.0047	3.3146	3.3168	3.0461	3.0000
J	5.8252e−04	1.7365e−05	8.8813e−04	3.6050e−05	4.1869e−11	**8.6711e−20**

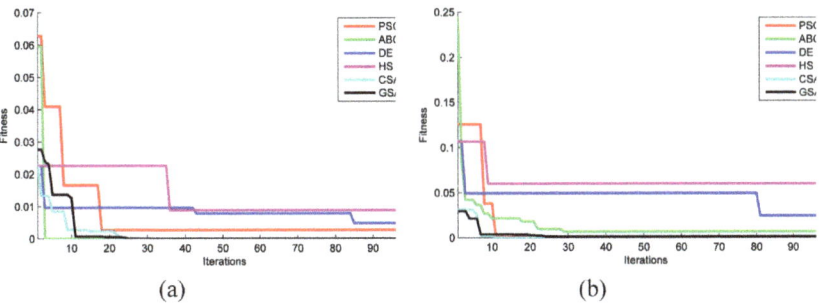

(a)   (b)

Fig. 3.3 Evolution of the objective function value for a two dimensional problem obtained by PSO, ABC, DE, HS, CSA and GSA for **a** Lorenz CS and **b** Chen CS

objective function values in the optimization process is shown in Fig. 3.3a for the Lorenz system and in Fig. 3.3b for the Chen CS. In Fig. 3.3a the ABC needs fewer iterations to converge towards the optimal value and in Fig. 3.3b such conduct occurs for the CSA, however, in both conditions, the GSA is the procedure that supplies more accurate fitness values.

3.4.1.3 Three-Dimensional Parameter Estimation

This investigation evaluates the full group of parameters for the Lorenz and Chen chaotic systems. In their characterization, they contain three parameters to be adjusted and in this examination, the selected algorithms consider the full set. This problem shows a high difficulty degree not only for the non-linearity of the CS but also for the dimensionality of the optimization problem. Equation 3.29 describes the three different parameters. The experimental results are shown in Table 3.5 which displays the calculated values for each parameter and the best fitness value acquired at the end of the optimization process.

Table 3.5 Performance results of three-dimensional problem for Lorenz system

	PSO	ABC	DE	HS	CSA	GSA
σ	11.0000	9.7413	10.0568	10.1445	9.9994	10.0100
r	28.9779	27.9440	27.9183	28.1141	28.0002	28.0002
b	3.0000	2.7191	2.6113	2.6112	2.6670	2.6631
J	4.0458e−04	5.2850e−04	0.0029	1.1198e−04	2.2072e−09	**1.3663e−16**

$$[\sigma, r, b] \tag{3.29}$$

Table 3.5 shows the results of the performance of PSO, ABC, DE, HS, CSA and GSA for the parameter analysis in the Lorenz CS. In this topic, the problem has three different parameters that must be adjusted. Comparable to the other tests each algorithm handles 100 generations to locate the optimal value of the cost function. The collection of parameters to be calculated by the evolutionary algorithms for the three-dimensional problem is expressed in Eq. 3.30.

$$[a, b, c] \tag{3.30}$$

In Table 3.6, the experimental outcomes acquired by the selected methods over the Chen CS in a three dimensional problem are presented. Regarding to the produced results by the algorithms, the GSA and CSA algorithms obtained better results regarding the rest of tested algorithms to be compared (Tables 3.5 and 3.6). Such fact corresponds to the intrinsic nature of the operators utilized for each method. GSA algorithm calculates better the parameters regarding to the total number of experiments. The evolution of the fitness through the iteration process is shown in Fig. 3.4, where it is feasible to study the convergence of the algorithms. In Fig. 3.4a the Lorenz CS is considered. According to the results, the methods that require fewer number of generations are ABC, CSA and GSA. On the other hand, Fig. 3.4b shows the results over the Chen system where it can shown that all methods tend to converge quickly. However, the ABC obtains its minimum considering a reduced number of generations, but the most accurate outcomes are calculated by the CSA method.

Table 3.6 Performance results of three-dimensional problem for Chen system

	PSO	ABC	DE	HS	CSA	GSA
a	30.7156	37.2663	31.9672	35.2433	35.0002	34.8441
b	26.1472	29.1574	26.3746	28.2414	28.0001	27.9170
c	4.0171	2.4549	3.3811	3.2948	2.9999	3.0189
J	0.0019	0.0049	0.0055	0.0124	**1.3222e−07**	3.1576e−05

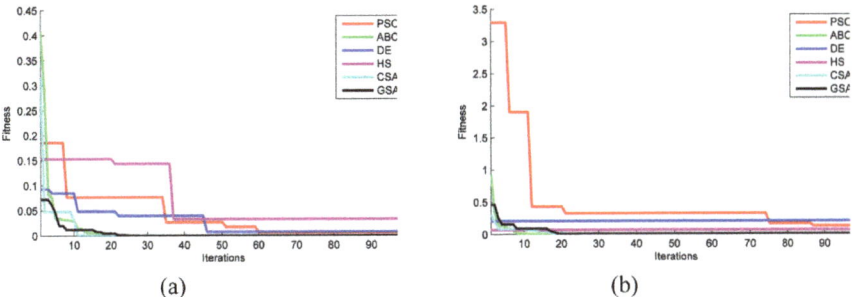

Fig. 3.4 Evolution of the objective function values for a three dimensional problem obtained by PSO, ABC, DE, HS, CSA and GSA for **a** Lorenz CS and **b** Chen CS

3.4.2 Statistical Analysis

To statistically validate the outcomes delivered by the six algorithms, a nonparametric statistical significance test is involved. This statistical test is known as Wilcoxon's rank sum test for independent samples [29, 30]. To make use of this rank sum test, 35 independent runs for each method were required to prove that there is a significant distinction among algorithms. However, because we have executed several Wilcoxon rank-sum tests, the probability to commit an error type 1 increases and we can abandon the null hypothesis even when it is true. To avoid this problem a correction for the p-value of Wilcoxon test is modified by the Bonferroni correction [31]. Merging the Wilcoxon test with an adjusted p-value from Bonferroni test, we can reveal that it exists a significant distinction among each pair of the methods. This examination is divided in two parts: a statistical analysis for Lorenz system and a statistical analysis for Chen system. Within these sections, the outcomes for each dimension are given in different tables. These tables include a column with the Bonferroni adjusted p-value employed to validate the Wilxocon test. If a cell within this column holds 1, the solutions provided by the couple of algorithms being approximated are significantly distinct among them.

3.4.2.1 Statistical Analysis for Lorenz System

A statistical study from the Lorenz chaotic system is achieved. In this investigation, the outcomes of each test given in Sect. 3.4.1 are supposed. In this reason, the outcomes are split in: (a) One-dimensional statistical analysis, (b) two-dimensional statistical analysis and (c) three-dimensional statistical analysis.

(a) *One-dimensional statistical analysis*

For this investigation, each Lorenz parameter is approximated for each method. The results for σ, r and b are presented in Table 3.7.

Table 3.7 Statistical results for the one-dimensional problem of estimate the σ, r and b parameters for Lorenz System

Versus	Lorenz CS (σ)		Lorenz CS (r)		Lorenz CS (b)	
	Wilcoxon (p)	Bonferroni (p)	Wilcoxon (p)	Bonferroni (p)	Wilcoxon (p)	Bonferroni (p)
PSO versus ABC	0.5183	0	0.0039	0	1.3529e−09	1
PSO versus CSA	6.5455e−13	1	6.5455e−13	1	6.5455e−13	1
PSO versus HS	0.001	1	0.3534	0	2.3362e−06	1
PSO versus GSA	6.5455e−13	1	6.5455e−13	1	6.5455e−13	1
PSO versus DE	2.2549e−10	1	0.2799	0	1.1798e−05	1
ABC versus CSA	6.5455e−13	1	6.5455e−13	1	6.5455e−13	1
ABC versus HS	4.5838e−05	1	0.0033	1	1.3349e−07	1
ABC versus GSA	6.5455e−13	1	6.5455e−13	1	6.5455e−13	1
ABC versus DE	9.5761e−12	1	3.9372e−05	1	1.2204e−08	1
CSA versus HS	6.5455e−13	1	6.5455e−13	1	6.5455e−13	1
CSA versus GSA	6.5455e−13	1	7.1329e−13	1	1.8196e−12	1
CSA versus DE	6.5455e−13	1	6.5455e−13	1	6.5455e−13	1
HS versus GSA	6.5455e−13	1	6.5455e−13	1	6.5455e−13	1
HS versus DE	4.3992e−07	1	0.8234	0	0.0042	0
GSA versus DE	6.5455e−13	1	6.5455e−13	1	6.5455e−13	1

(b) *Two-dimensional statistical analysis*

In this experimentation, each couple of parameters (given in Eq. 3.27) are studied. The outcomes of the Wilcoxon rank-sum test and the Bonferroni correction for the couples of elements $[\sigma, r]$, $[\sigma, b]$ and $[r, b]$ are presented in Table 3.8.

(c) *Three-dimensional statistical analysis*

In this experimentation, all the parameters $[\sigma, r, b]$ of the Lorenz chaotic system are obtained. The outcomes of the test of Wilcoxon and Bonferroni for this three-dimensional problem are presented in Table 3.9.

3.4.2.2 Statistical Analysis for Chen System

This section gives the statistical examination of the outcomes acquired from the Chen CS. This study consists of three parts: (a) One-dimensional, (b) Two-dimensional and (c) Three-dimensional statistical study. They are employed to confirm the values accepted in each experiment from the Sect. 3.4.1.

(a) *One-dimensional statistical analysis*

For this investigation, each parameter of the Chen CS is sampled considering all the algorithms. The values of the Wilcoxon test and Bonferroni correction for the Chen parameters are individually presented in Table 3.10.

(b) *Two-dimensional statistical analysis*

This section gives the experimental outcomes over the data acquired for the two-dimensional test using the Chen chaotic systems. This investigation studies the cost function values of all the six ECT in the estimate of each couple of parameters of Eq. 3.28, the values are shown in Table 3.11.

(c) *Three-dimensional statistical analysis*

For the investigations of this section, all the parameters $[a, b, c]$ of the Chen CS are evaluated. The Bonferroni correction and the Wilcoxon rank test are involved regarding the three-dimensional problem defined in Eq. 3.30. The acquired values that verify the performance of the algorithms are indicated in Table 3.12.

The obtained outcomes by the Wilcoxon test and the Bonferroni correction are exhibited in Tables 3.7, 3.8, 3.9, 3.10, 3.11 and 3.12. There are 15 comparisons among methods for problems with distinct dimensions. For the case of Wilcoxon test, all p values registered in the tables are estimated with 0.05 (5% significance level) which is strong evidence against the null hypothesis. Nevertheless, the number of executions that the Wilcoxon test requires is big, the possibility of making one or more Type I errors in a family of comparisons increases. To construct valid statistical results for numerous comparison problems, the correction for the p value will

Table 3.8 Statistical results for the two-dimensional problem of estimate the σ, r and b parameters for Lorenz System

Versus	Lorenz CS (σ, r)		Lorenz CS (σ, b)		Lorenz CS (r, b)	
	Wilcoxon (p)	Bonferroni (p)	Wilcoxon (p)	Bonferroni (p)	Wilcoxon (p)	Bonferroni (p)
PSO versus ABC	0.0071	0	0.2004	0	1.4622e−05	1
PSO versus CSA	6.5455e−13	1	6.5455e−13	1	6.5455e−13	1
PSO versus HS	0.0085	0	0.9813	0	8.5166e−06	1
PSO versus GSA	1.1741e−07	1	6.5455e−13	1	6.5455e−13	1
PSO versus DE	0.0108	0	0.0272	1	3.7218e−04	1
ABC versus CSA	6.5455e−13	1	6.5455e−13	1	6.5455e−13	1
ABC versus HS	0.3847	0	0.3238	0	0.8786	0
ABC versus GSA	4.8024e−10	1	6.5455e−13	1	6.5455e−−13	1
ABC versus DE	0.2695	0	0.4109	0	0.0821	0
CSA versus HS	6.5455e−13	1	6.5455e−13	1	6.5455e−13	1
CSA versus GSA	2.5964e−09	1	6.5455e−13	1	6.5455e−13	1
CSA versus DE	6.5455e−13	1	6.5455e−13	1	6.5455e−13	1
HS versus GSA	1.6602e−10	1	6.5455e−13	1	6.5455e−13	1
HS versus DE	0.9158	0	0.1155	0	0.0485	0
GSA versus DE	6.0469e−11	1	6.5455e−13	1	6.5455e−13	1

Table 3.9 Statistical results for the three-dimensional problem of estimate the σ, r and b parameters for Lorenz System

Versus	Lorenz CS (σ, r, b)	
	Wilcoxon (p)	Bonferroni (p)
PSO versus ABC	0.0136	0
PSO versus CSA	6.5455e−13	1
PSO versus HS	5.6036e−05	1
PSO versus GSA	1.4118e−12	1
PSO versus DE	0.0526	0
ABC versus CSA	6.5455e−13	1
ABC versus HS	0.0345	0
ABC versus GSA	2.7700e−12	1
ABC versus DE	0.3068	0
CSA versus HS	6.5455e−13	1
CSA versus GSA	1.2855e−04	1
CSA versus DE	6.5455e−13	1
HS versus GSA	2.3326e−11	1
HS versus DE	6.5834e−04	1
GSA versus DE	6.5455e−13	1

eliminate the presence of false positives [32, 33]. The approach used in this work is the well-known Bonferroni correction. According to the statistical outcomes, only the solutions produced by GSA and CSA algorithms are significantly distinct respect each other and even distinct considering the rest of the evolutionary algorithms. These distinctions are displayed for every dimensional problem for each chaotic system that has been employed.

3.5 Conclusions

This chapter shows a comparison analysis among six optimization techniques for parameter estimation for chaotic systems problems. Under this investigation, parameter estimation is assumed as an optimization problem. In the comparison, particular attention is produced to recently developed evolutionary methods such as the Gravitational Search Algorithm (GSA) and Cuckoo Search (CSA), also containing well-known approaches such as the Particle Swarm Optimization (PSO), the Artificial Bee Colony optimization (ABC), Harmony Search (HS) and Differential Evolution (DE).

The comparison has been experimentally estimated over each dimension for each CS. The experimentation results have indicated that GSA and CSA outperform PSO, ABC, HS, DE in terms of accurateness and a statistically significant framework (Bonferroni correction).

Table 3.10 Statistical results for the one-dimensional problem of estimate the a, b and c parameters for Chen System

Versus	Chen CS (a)		Chen CS (b)		Chen CS (c)	
	Wilcoxon (p)	Bonferroni (p)	Wilcoxon (p)	Bonferroni (p)	Wilcoxon (p)	Bonferroni (p)
PSO versus ABC	3.7218e−04	1	0.0026	1	0.0316	0
PSO versus CSA	6.5455e−13	1	6.5455e−13	1	6.5455e−13	1
PSO versus HS	0.557	0	0.8234	0	0.1519	0
PSO versus GSA	6.5455e−13	1	6.5455e−13	1	6.5455e−13	1
PSO versus DE	6.1187e−06	1	0.4043	0	1.1741e−07	1
ABC versus CSA	6.5455e−13	1	6.5455e−13	1	6.5455e−13	1
ABC versus HS	0.0512	0	0.0023	1	5.5367e−04	1
ABC versus GSA	6.5455e−13	1	6.5455e−13	1	6.5455e−13	1
ABC versus DE	1.1288e−10	1	2.3643e−04	1	8.0537e−09	1
CSA versus HS	6.5455e−13	1	6.5455e−13	1	6.5455e−13	1
CSA versus GSA	7.1329e−13	1	6.5455e−13	1	7.7721e−13	1
CSA versus DE	6.5455e−13	1	6.5455e−13	1	6.5455e−13	1
HS versus GSA	6.5455e−13	1	6.5455e−13	1	6.5455e−13	1
HS versus DE	0.0062	0	0.2747	0	9.11e−07	1
GSA versus DE	6.5455e−13	1	6.5455e−13	1	6.5455e−13	1

Table 3.11 Statistical results for the two-dimensional problem of estimate the a, b and c parameters for Chen System

Versus	Chen CS (a, b)		Chen CS (b, c)		Chen CS (a, c)	
	Wilcoxon (p)	Bonferroni (p)	Wilcoxon (p)	Bonferroni (p)	Wilcoxon (p)	Bonferroni (p)
PSO versus ABC	0.0054	0	0.4313	0	0.6724	0
PSO versus CSA	6.5455e−13	1	6.5455e−13	1	6.5455e−13	1
PSO versus HS	0.0101	0	0.6135	0	0.2355	0
PSO versus GSA	1.1569e−06	1	1.9797e−12	1	7.7721e−13	1
PSO versus DE	0.0052	0	0.0022	1	0.0355	0
ABC versus CSA	6.5455e−13	1	6.5455e−13	1	6.5455e−13	1
ABC versus HS	0.9251	0	0.769	0	0.557	0
ABC versus GSA	3.4569e−09	1	1.9797e−12	1	1.0046e−12	1
ABC versus DE	0.9345	0	0.0801	0	0.2219	0
CSA versus HS	6.5455e−13	1	6.5455e−13	1	6.5455e−13	1
CSA versus GSA	6.5455e−13	1	1.9354e−10	1	1.2228e−11	1
CSA versus DE	6.5455e−13	1	6.5455e−13	1	6.5455e−13	1
HS versus GSA	1.3073e−08	1	2.1535e−12	1	1.0046e−12	1
HS versus DE	1	0	0.0355	0	0.5729	0
GSA versus DE	2.0919e−09	1	1.0046e−12	1	6.5455e−13	1

Table 3.12 Statistical results for the three-dimensional problem of estimate the a, b and c parameters for Chen System

Versus	Chen CS (a, b, c)	
	Wilcoxon (p)	Bonferroni (p)
PSO versus ABC	0.8694	0
PSO versus CSA	6.5455e−13	1
PSO versus HS	0.1025	0
PSO versus GSA	1.2204e−08	1
PSO versus DE	0.4109	0
ABC versus CSA	6.5455e−13	1
ABC versus HS	0.0761	0
ABC versus GSA	1.4002e−08	1
ABC versus DE	0.4109	0
CSA versus HS	6.5455e−13	1
CSA versus GSA	6.5455e−13	1
CSA versus DE	6.5455e−13	1
HS versus GSA	1.9354e−10	1
HS versus DE	0.3783	0
GSA versus DE	8.0537e−09	1

The remarkable performance of GSA and CSA is described by the evolutionary mechanisms that permit better exploration of the search space, raising the capacity to locate multiple optima and their exploitation mechanisms.

References

1. Zhou X, Yang C, Gui W (2014) Nonlinear system identification and control using state transition algorithm. Appl Math Comput 226:169–179
2. Albaghdadi M, Briley B, Evens M (2006) Event storm detection and identification in communication systems. Reliab Eng Syst Saf 91:602–613
3. FrankPai P, Nguyen B-A, Sundaresan MJ (2013) Nonlinearity identification by time-domain-only signal processing. Int J Non-Linear Mech 54:85–98
4. Yang SK, Chen CL, Yau HT (2002) Control of chaos in Lorenz system. Chaos Solitons Fractals 13(4):767–780
5. Yassen MT (2003) Chaos control of Chen chaotic dynamical system. Chaos Solitons Fractals 15(2):271–283
6. Na J, Ren X, Xia Y (2014) Adaptive parameter identification of linear SISO systems with unknown time-delay. Syst Control Lett 66:43–50
7. Dai C, Chen W, Zhu Y (2010) Seeker optimization algorithm for digital IIR filter design. IEEE Trans Ind Electron 57:1710–1718
8. He Q, Wang L, Liu B (2007) Parameter estimation for chaotic systems by particle swarm optimization. Chaos Solitons Fractals 34(2):654–661
9. Kennedy J, Eberhart RC (1995) Particle swarm optimization. In: Proceedings of IEEE international conference neural networks, vol 4, pp 1942–1948

10. Karaboga D (2005) An idea based on honey bee swarm for numerical optimization. Technical report TR06, Comput. Eng. Dep. Eng. Fac. Erciyes Univ.
11. Geem ZW, Kim JH, Loganathan GV (2001) A new heuristic optimization algorithm: harmony search. SIMULATION 762:60–68
12. Storn R, Price K (1997) Differential evolution–a simple and efficient heuristic for global optimization over continuous spaces. J Glob Optim 341–359
13. Yang X-S, Deb S (2010) Cuckoo search via Levy flights, pp 210–214
14. Rashedi E, Nezamabadi-pour H, Saryazdi S (2009) GSA: a gravitational search algorithm. Inf Sci (NY) 179(13):2232–2248
15. Peng B, Liu B, Zhang FY, Wang L (2009) Differential evolution algorithm-based parameter estimation for chaotic systems. Chaos Solitons Fractals 39(5):2110–2118
16. Cuevas E, Gálvez J, Hinojosa S, Avalos O, Zaldívar D, Pérez-cisneros M (2014) A comparison of evolutionary computation techniques for IIR model identification 2014
17. Ahn C (2006) Advances in evolutionary algorithms: theory, design and practice. Springer
18. Chiong R, Weise T, Michalewicz Z (2012) Variants of evolutionary algorithms for real-world applications. Springer
19. Oltean M (2007) Evolving evolutionary algorithms with patterns. Soft Comput 11:503–518
20. Lin Y-L, Chang W-D, Hsieh J-G (2008) A particle swarm optimization approach to nonlinear rational filter modeling. Expert Syst Appl 34:1194–1199
21. Yang X-S, Deb S (2009) Cuckoo search via Lévy flights. In: Proceedings of world congress on nature and biologically inspired computing (NABIC '09), pp 210–214
22. Pavlyukevich I (2007) Lévy flights, non-local search and simulated annealing. J Comput Phys 226:1830–1844
23. Mantegna RN (2007) Fast, accurate algorithm for numerical simulation of Lévy stable stochastic processes. Phys Rev E 49:4677–4683
24. Storn R, Price K (1997) Differential evolution—a simple and efficient heuristic for global optimization over continuous spaces. J Glob Optim
25. De Jong K (1988) Learning with genetic algorithms: an overview. Mach Learn 3:121–138
26. Hegazi AS, Ahmed E, Matouk AE (2011) The effect of fractional order on synchronization of two fractional order chaotic and hyperchaotic systems. J Fract Calc Appl 1:1–15
27. Huang L, Feng R, Wang M (2004) Synchronization of chaotic systems via nonlinear control. Phys Lett A 320(4):271–275
28. Patwardhan, Patidar R, George NV. On a cuckoo search optimization approach towards feedback system identification
29. García S, Molina D, Lozano M, Herrera F (2009) A study on the use of non-parametric tests for analyzing the evolutionary algorithms. J Heuristics 15:617–644
30. Shilane D, Martikainen J, Dudoit S, Ovaska SJ (2008) A general framework for statistical performance comparison of evolutionary computation algorithms. Inf Sci (NY) 178:2870–2879
31. Bonferroni CE (1935) Il calcolo delle assicurazioni su gruppi di teste. Stud. Onore del Profr. Salvatore Ortu Carboni 13–60
32. Cuevas E, González A, Fausto F, Zaldívar D, Pérez-Cisneros M (2015) Multithreshold segmentation by using an algorithm based on the behavior of locust swarms. Math Probl Eng 2015:805357
33. Hinojosa S, Oliva D, Cuevas E, Avalos O, Gálvez J et al (2018) Improving multi-criterion optimization with chaos: a novel multi-objective chaotic crow search algorithm, neural computing and applications. 29(8):319–335

Chapter 4
Comparison Study of Novel Evolutionary Algorithms for Elliptical Shapes in Images

Shape recognition in digital image processing describes one of the difficult and hard-solving situations in artificial vision due to its nonlinear and stochastic structure. Traditional image processing methods have been commonly employed to solve this situation. Additionally, shape recognition considers evolutionary computation techniques. They have been exposed to better performance in terms of accurateness than traditional optimization methods. Traditionally, many evolutionary schemes employ old search mechanisms to determine geometrical designs within images. Original optimization methods experiment with many constraints such as low diversity and premature convergence. Since the development of novel evolutionary structures, new development evolutionary methods have overcome these complexities to obtain better results. This chapter presents a comparative study based on the use of five of the most recent evolutionary algorithms to increase the performance of the shape recognition problem. Such methods as the Cuckoo Search (CS), Gravitational Search Algorithm (GSA), Crow Search Algorithm (CSA), Grey Wolf Optimizer (GWO), and Whale Optimizer Algorithm (WOA). Since these methods have been employed in many applications, the main goal is to decide their accuracy when they confront a complex problem such as shape recognition. Numerical results, conducted over a set of experiments comprised images with distinct complexity levels.

4.1 Introduction

The goal of shape detection is to identify geometric shapes such as circles, ellipses, and lines. Shape recognition emerges in many domains of pattern recognition according to several industrial applications such as autonomous navigation [1]. Shape recognition is generally performed using the well-known method called Hough Transform (HT) [1, 2]. A standard version of the Hough-based approach considers the information of the edges obtained by an edge identification algorithm to acquire prospect geometrical forms. In the procedure, each edge pixel, which overlaps against

© The Author(s), under exclusive license to Springer Nature Switzerland AG 2023
E. Cuevas et al., *Analysis and Comparison of Metaheuristics*, Studies in Computational Intelligence 1063, https://doi.org/10.1007/978-3-031-20105-9_4

a speculatory geometrical form, votes within the parameter domain. Then, the most likely geometric shape is acquired by averaging, filtering, and histogramming the parameter space. HT is a strong algorithm; nonetheless, it holds some important drawbacks. HT needs a large quantity of memory space to allocate all assumed parameters of the geometrical figure. It also requires a high computational complexity, which causes poor performance in terms of speed. Another critical situation is based on accurateness. In HT, the preciseness of the acquired parameters for the detected form is decreased, particularly in presence of noise [3].

Alternatively to the HT method, the problem of shape recognition has also been solved via evolutionary algorithms. In general terms, they have demonstrated better results than the results obtained by traditional HT [4]. A recognition strategy, based on an evolutionary method, employs a group of edge points as decision variables to define speculative shapes. A matching function evaluates if a speculatory geometric shape is contained within the image. Thus, operated by the values of the matching function, the group of encoded points is operated considering evolutionary approaches in which the best solutions define the original geometric forms. Several evolutionary algorithms have been employed to obtain many interesting shape detectors such as Genetic algorithms (GA) [5, 6] and Particle Swarm Optimization (PSO) [7], Differential Evolution (DE) [8], Cloning Selection method (CSM) [9], Harmony Search (HS) [10], Artificial Bee Colony (ABC) [11] and Animal Behavior (AB) [12, 13].

Lately, evolutionary methods have been attracted the attention of several communities [14]. Such methods represent challenging opposition to traditional optimization methods, which are prone to fail in multiple simple problems. On the other hand, initial evolutionary methods maintain in their structure many drawbacks such as premature convergence and population diversity [15]. Novel evolutionary methods have handled these problems providing better results [16]. Many of these new evolutionary methods have also been introduced lately. In general, they suggest new standards based on evolutionary operators to obtain acceptable exploration and exploitation stages inside large search spaces regarding a considerable number of dimensions. This renewed generation of evolutionary methods is extensive, some examples include Crow Search Algorithm (CSA) [17], Cuckoo Search (CS) [18], Moth-flame Optimization Algorithm (MFOA) [19], Grey Wolf Optimizer (GWO) [20], Whale Optimization Algorithm (WOA) [21], Gravitational Search Algorithm (GSA) [22], Sine Cosine Algorithm (SCA) [23], Cricket Behaviour-based Evolutionary (CBE) [24], Dragonfly Algorithm (DF) [25], Hybrid Cognitive Optimization Algorithm (HCOA) [26], Interactive Search Algorithm (ISA) [27], Farmland Fertility Algorithm (FFA) [28], Binary Artificial Algae Algorithm (BAAA) [29], Electro-Search algorithm (ESA) [30], and Island Bat Algorithm for Optimization (IBAO) [31].

A review of the related literature, there is no existence of a single optimization method which can solve the entire set of optimization problems in an effective way. Evolutionary techniques have been designed considering important aspects to efficiently solve specific situations. For that, to evaluate the performance of a given optimization procedure, its performance is evaluated within the context of the optimization problem. Multiple comparison studies between evolutionary methodologies have been introduced in the literature in many fields.

This chapter introduces a comparative study on the application of novel and famous evolutionary methods to the shape identification problem such as the Cuckoo Search (CS) Grey Wolf Optimizer (GWO), Gravitational Search Algorithm (GSA), Crow Search Algorithm (CSA), and Whale Optimizer Algorithm (WOA). These methods have been tested in many research fields. For this, the main goal is to compute their efficiency when they confront complex situations such as shape recognition. Such techniques have been picked for this study, since they are considered the most popular methods over the last five years. The optimization problem of shape recognition concerns the recognition of circles, circular arcs, ellipses, and lines. In this chapter, in order to present an interesting comparison among all the techniques, the ellipse form detection problem has been selected. It is considered one the most complex recognition problems. The comparison is split into two parts. In the first part, the performance of all evolutionary methods is evaluated regarding the ellipse identification over natural images, while, in the second part, the scalability of the detection for each method in the presence of complex scenarios is investigated. In the analysis, many performance metrics have been considered to estimate the accurateness Further, test scenarios are conducted over the experimental data in order to validate the outcomes.

The remainder of this chapter is organized as: In Sect. 4.2 describes the shape recognition problem formulation such as an optimization task; in Sect. 4.3 the main element of each evolutionary approach are shown; in Sect. 4.4, a comparative viewpoint of the evolutionary mechanisms is presented; in Sect. 4.5, the numerical simulations are shown. Finally, in Sect. 4.6, some conclusions are given.

4.2 Problem Definition

In order to recognize ellipses within digital images, they must be preprocessed based on an edge detection technique that delivers an edge map. Then, the location (x_i, y_i) of each edge e_i is contained in a vector of edges $E = \{e_1, e_2, \ldots, e_{N_p}\}$; where N_p represents the total amount of localized edge pixels. Table 4.1 shows a description of each symbol used in this section.

Under this methodology, a candidate ellipse C is coded based on five distinct edge points within its perimeter. Hence, candidate ellipses are constructed by randomly choosing edge points from the vector of edges E. Figure 4.1 illustrates the implemented process.

Within this scheme, following Fig. 4.1, a candidate ellipse is characterized as the ellipse which passes through the five following points e_p, e_q, e_r, e_s, and e_t $C = \{e_p, e_q, e_r, e_s, e_t\}$. Then, an ellipse is mathematical defined as follows:

Table 4.1 Representation of the set of symbol used in this chapter

(x_i, y_i)	The position of the pixel
$E = \{e_1, e_2, \ldots, e_{N_p}\}$	The vector of edges
N_p	The amount of detected edge pixels
$C = \{e_p, e_q, e_r, e_s, e_t\}$	Candidate ellipse coded base on five edge points
$\hat{a}, \hat{b}, \hat{c}, \hat{f}, \hat{g}$ and $\hat{h}$	Ellipse coefficients
a, b, f, g and h	Normalized coefficients where $a = \hat{a}/\hat{c}, b = \hat{b}/\hat{c}, f = \hat{f}/\hat{c},$ $g = \hat{g}/\hat{c}$ and $h = \hat{h}/\hat{c}$
(x_0, y_0)	Ellipse center
r_{max}, r_{min}	Maximal radius and minimal radii, respectively
θ	The orientation of the ellipse
R and D	Temporary variables $R = \sqrt{(a - b)^2 + 4h^2}$ $D = ab - h^2$
Δ	The determinant of the parameters of the ellipse $\det \begin{vmatrix} a & h & g \\ h & b & f \\ g & f & 1 \end{vmatrix}$
$P = \{p_1, p_2, \ldots, p_{N_e}\}$	Candidate set of points for the ellipse
N_e	Number of candidate points for the ellipse
$p_w = (x_w, y_w)$	Pixel position $w \in 1, \ldots, N_e$
$J(C)$	Error of the estimated ellipse
$f_C(p_w)$	This function corresponds to verify if p_w is inside or outside the perimeter of the ellipse
$Q(p_w)$	This function verifies the existence of the perimeter point p_w inside the vector of edges

Fig. 4.1 Candidate elliptical
shape solution

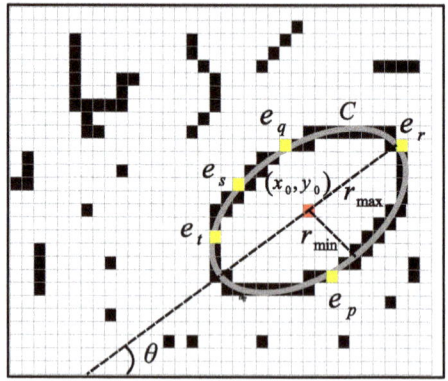

$$\hat{a}x_i^2 + 2\hat{h}x_i\,y_i + \hat{b}y_i^2 + 2\hat{g}x_i + 2\hat{f}y_i + \hat{c} = 0 \tag{4.1}$$

Therefore, substituting the coordinates of each parameter e_k $(k \in p, q, r, s, t)$ of C into Eq. 4.1, a set of five equations is defined. Therefore, after determining the values $\hat{a}, \hat{b}, \hat{c}, \hat{f}$, and $\hat{g}$, and dividing by the constant $\hat{c}$, the mathematical description is as follows:

$$ax_i^2 + 2hx_i\,y_i + by_i^2 + 2gx_i + 2fy_i + 1 = 0 \tag{4.2}$$

From this mathematical model, the global geometrical features of the ellipse such as the, orientation θ, maximal radius r_{max}, minimal radius r_{min}, and the center (x_0, y_0) can be computed as follows:

$$x_0 = \frac{hf - bg}{D}, \; y_0 = \frac{gh - af}{D} \tag{4.3}$$

$$r_{max} = \sqrt{\frac{-2\Delta}{D(a + b - R)}}, \; r_{min} = \sqrt{\frac{-2\Delta}{D(a + b + R)}} \tag{4.4}$$

$$\theta = \frac{1}{2}\arctan\left(\frac{2h}{a - b}\right), \tag{4.5}$$

where

$$R = \sqrt{(a - b)^2 + 4h^2}, \, D = ab - h^2, \, \Delta = \det\begin{vmatrix} a & h & g \\ h & b & f \\ g & f & 1 \end{vmatrix} \tag{4.6}$$

To compute the error function $J(C)$ of each ellipse C, the group P of its coordinate points is obtained. This group P contains the N_e pixel points $\{p_1, p_2, \ldots, p_{N_e}\}$ that the ellipse could incorporate. Then, it is verified, if such points are contained against the edge map.

The group P is obtained by employing the Midpoint Ellipse Algorithm (MEA) [32] which is a plotting method that locates the necessary points $\{p_1, p_2, \ldots, p_{N_e}\}$ to draw an elliptical figure. Each pixel coordinate $p_w = (x_w, y_w)$ of the perimeter must fulfill the following association $f_C(p_w) = r_{max}x_w^2 + r_{min}y_w^2 - r_{max}^2 r_{min}^2$. Under this condition, the MEA technique conducts the test to validate the position of every pixel point to determine if it is discovered inside or outside the perimeter of the ellipse. Thus, if the pixel p_w is contained inside the ellipse perimeter, the ellipse function $f_C(p_w)$ obtains a negative value. On the other hand, if the pixel p_w is outside the perimeter of the ellipse, the ellipse function $f_C(p_w)$ produces a positive value. Under this scheme, the MEA algorithm discovers the pixel coordinates $\{p_1, p_2, \ldots, p_{N_e}\}$ that yields values from $f_C(\cdot)$ close to zero.

Once defined the points of the perimeter P of the candidate ellipse, the error function $J(C)$ is calculated to measure the matching error. This value can be represented as the cost function to be optimized. $J(C)$ represents the matching similarity among the perimeter and the pixels that corresponds within the edge map. Hence, $J(C)$ is defines according to:

$$J(C) = 1 - \frac{\sum_{u=1}^{N_e} Q(p_w)}{N_e}, \tag{4.7}$$

where $Q(p_w)$ corresponds to a relation model that evaluates the existence of the perimeter inside the vector od edges E. This mechanism can be represented as follows:

$$Q(p_w) = \begin{cases} 1 \text{ if } p_w \in E \\ 0 \text{ otherwise} \end{cases}, \tag{4.8}$$

If the value evaluated by $J(C)$ is close to zero, then it implies a better estimation of the ellipse shape regarding the ellipse contained in the original image. Figure 4.2 shows the procedure over the evaluation of the matching error function. Figure 4.2a presents the edge map of the image. Assuming that the ellipse is grouped by the edge pixels e_p, e_q, e_r, e_s and e_t from the vector of edges, Fig. 4.2b illustrates the coordinates of the perimeter $P = \{p_1, p_2, \ldots, p_{N_e}\}$ obtained by the MEA algorithm. In the figure, it illustrates an ellipse of 52-pixel points. Finally, in Fig. 4.2c, the set of points of the parameter P is overlapped against the edge map point by point, to identify the concurrencies among P and E. In Fig. 4.2c, just 29 coordinate points are shown for both vectors P and E. Considering these values, the matching error function $J(C)$ is 0.4423.

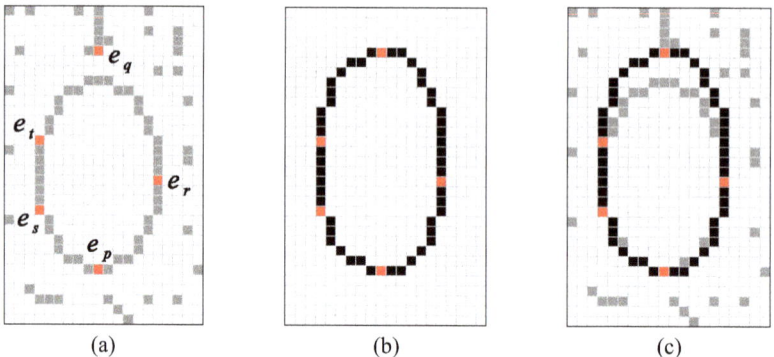

(a)   (b)   (c)

Fig. 4.2 Cost fitness function $J(C)$. **a** Represents the edge map. **b** Represents the coordinates of the perimeter. And, Fig. 4.2c shows the concurrencies between (**a**) and (**b**)

4.2.1 Multiple Ellipse Detection

To identify multiple samples of elliptical shapes, the proposed methodology holds the best ellipse candidate been so far by each generation k based on memory structure $\mathbf{M}(k)$. Hence, once the evolutionary algorithm has been finalized, a depuration procedure inside the memory structure is achieved. The main goal is to consider which ellipses in $\mathbf{M}(k)$ represents an ellipse according to the edge map. Assuming that many elements from $\mathbf{M}(k)$ can define the same elliptical shape, a distinctiveness variable D_{S_1,S_2} is required to compute the discrepancy among two given elliptical shapes: S_1 and S_2. D_{S_1,S_2} is defined considering the Eq. 4.9 as:

$$D_{S_1,S_2} = \left|x_0^{S_1} - x_0^{S_2}\right| + \left|y_0^{S_1} - y_0^{S_2}\right| + \left|r_{max}^{S_1} - r_{max}^{S_2}\right| + \left|r_{min}^{S_1} - r_{min}^{S_2}\right| + \left|\theta_{S_1} - \theta_{S_2}\right|$$

(4.9)

Once calculated the distinctiveness variable D_{S_1,S_2}, a threshold value Th is also determined to evaluate if S_1 and S_2 can be considered as similar or not as follows:

$$Th = \frac{\left(\left|r_{max}^h - r_{max}^l\right| + \left|r_{min}^h - r_{min}^l\right|\right)}{4},$$

(4.10)

where $\left[r_{max}^l, r_{max}^h\right]$ and $\left[r_{min}^l, r_{min}^h\right]$ represents the possible ranges for the maximal and minimal radius r_{max} and r_{min}, respectively. Considering this approach, the elliptical shapes S_1 and S_2 define similar shapes, if $D_{S_1,S_2} < Th$, otherwise, they are considered different elliptical shapes. Hence, each ellipse from $\mathbf{M}(k)$ is compared among each memory element. The main goal is to discover the elliptical shapes based on the evaluation of the cost function $J(\cdot)$.

4.3 Evolutionary Optimization Techniques

Optimization algorithms have been developed to find the global solution Gs of a complex non-linear problem considering box constraints under the following model [33]:

$$Gs = \underset{\mathbf{x}=(x_1,\ldots,x_n)\in\mathbf{X}}{\arg\min} \quad J(\mathbf{x})$$

(4.11)

where $J(\mathbf{x})$ corresponds to a non-linear cost function while $\mathbf{X} = \{\mathbf{x} \in \mathbb{R}|l_i \leq x_i \leq u_i, i = 1, \ldots, n\}$ represents a constrained search space, restricted by the lower (l_i) and upper (u_i) bounds.

To find the optimal solution for Eq. 4.11, from an evolutionary computation point of view [33], a set of individuals $\mathbf{P}(k)$ ($\{\mathbf{p}_1(k), \mathbf{p}_2(k), \ldots, \mathbf{p}_{N_i}(k)\}$) of N_i possible solutions are evolved from an initial iteration (where $k = 0$) to a maximum number of generations (where $k = Maxgen$). As an initial phase, an evolutionary methodology obtains a set of N_i possible solutions whose fitness values are chosen based on an uniform distribution among the bounds lower (l_i)) and upper (u_i). At each generation k, a collection of evolutionary operators is employed to obtain a new population $\mathbf{P}(k + 1)$ from the original $\mathbf{P}(k)$. Each possible solution $\mathbf{p}_i(k)$ within the population $\mathbf{P}(k)$ corresponds to an n-dimensional decision variable vector $\{p_{i,1}(k), p_{i,2}(k), \ldots, p_{i,n}(k)\}$. The decision variable vector coded for each individual $\mathbf{p}_i(k)$ is evaluated using its values into the cost function $J(\mathbf{p}_i(k))$. Along the entire optimization (evolutionary) process, the best solution $\mathbf{b}$ been seen so far is stored and maintained.

In this section, it is described the principal features of five of the most popular evolutionary techniques to evaluate their performance by solving the elliptical shape recognition optimization problem. Such methodologies includes the Cuckoo Search (CS), Gravitational Search Algorithm (GSA), Whale Optimizer Algorithm (WOA), Crow Search Algorithm (CSA), and Grey Wolf Optimizer (GWO).

4.3.1 Grey Wolf Optimizer (GWO) Algorithm

The Grey Wolf Optimizer (GWO) [20] is a novel optimization algorithm based on the hunting process and the social hierarchy of the wolves.

4.3.1.1 Hierarchy

To computationally implement the abstracted model of the social hierarchy of the wolves, the best solution $\mathbf{b}$ is represented by $\mathbf{p}_\alpha$. The two adjacent solutions from the best one are represented by $\mathbf{p}_\beta$ and $\mathbf{p}_\delta$, respectively. The symbols ($\alpha, \beta, \delta \in \mathbf{P}(k)$). The rest of individuals are represented by ω wolves ($\omega \in \mathbf{P}(k), \omega \neq \alpha, \beta, \delta$). Hence, through the optimization process, the mechanism is guided by the $\mathbf{p}_\alpha$, $\mathbf{p}_\beta$, and $\mathbf{p}_\delta$ solutions while the remaining $\mathbf{p}_\omega$ wolves follow them.

4.3.1.2 Covering the Prey

Considering the behavior of the Grey wolves, the wolves try to encircle their prey. This procedure is defined as:

$$\mathbf{d} = \left| C\mathbf{p}_p(k) - \mathbf{p}_\omega(k) \right| \tag{4.12}$$

$$\mathbf{p}_\omega(k + 1) = \mathbf{p}_p(k) - A\mathbf{d}, \tag{4.13}$$

where k corresponds the actual iteration (generation), A and C corresponds two fixed coefficients. $\mathbf{p}_w$ and $\mathbf{p}_p$ code the position of the wolf and prey, respectively. In the Eqs. 4.10 and 4.11, the values of A and C are computed according to:

$$A = 2ar_1 - 1 \tag{4.14}$$

$$C = 2r_2, \tag{4.15}$$

where a is a linear decreased parameter which ranges from 2 to 0 along the optimization process. Additionally, r_1 and r_2 correspond to two random numbers from the range $[0, 1]$.

4.3.1.3 Hunting

Considering the hunting behavior of wolves, the search mechanism is guided principally by the Alfa wolf $\mathbf{p}_\alpha$ and sometimes by the wolves $\mathbf{p}_\beta$ and the $\mathbf{p}_\delta$. Under this mechanism, the remaining $\mathbf{p}_\omega$ wolves adjust their positions considering the positions of these wolves. This behavior is computed as:

$$\mathbf{d}_\alpha = |C_1\mathbf{p}_\alpha(k) - \mathbf{p}_\omega(k)|, \, \mathbf{d}_\beta = \big|C_2\mathbf{p}_\beta(k) - \mathbf{p}_\omega(k)\big|, \, \mathbf{d}_\delta = |C_3\mathbf{p}_\delta(k) - \mathbf{p}_\omega(k)| \tag{4.16}$$

$$L_1 = \mathbf{p}_\alpha(k) - A_1\mathbf{d}_\alpha, \, L_2 = \mathbf{p}_\beta(k) - A_2\mathbf{d}_\beta, \, L_3 = \mathbf{p}_\delta(k) - A_3\mathbf{d}_\delta, \tag{4.17}$$

where A_1, A_2, A_3, C_1, C_2 and C_3 are fixed coefficients calculated based on Eqs. 4.12 and 4.13. Once calculated, the location in the following iteration $(k + 1)$ of each wolf is modified as:

$$\mathbf{p}_\omega(k + 1) = (L_1 + L_2 + L_3)/3, \tag{4.18}$$

4.3.1.4 Exploration and Exploitation

Both strategies are coded in the search mechanism considering the value of a in Eq. 4.12. While a diminishes its value, the variation of A also represents a reduction effect. Hence, when $|A| > 1$ the procedure explores the search space, while $|A| < 1$ the searching procedure centralizes the solutions over the global solution been so far.

4.3.2 Whale Optimizer Algorithm (WOA)

The Whale Optimizer Algorithm (WOA) [21] is an fascinating evolutionary technique which is based on the spiral bubble-net hunting behavior in humpback whales.

4.3.2.1 Encircling a Prey

In the WOA mechanism, the best possible solution **b** seen so-far is represented by $\mathbf{p}_b$. Hence, the rest humpback whales $\mathbf{p}_G$ ($\mathbf{p}_G \in \mathbf{P}(k)$) will be attracted by such individual. This mechanism is defined as:

$$\mathbf{s} = |C\mathbf{p}_b(k) - \mathbf{p}_G(k)| \tag{4.19}$$

$$\mathbf{p}_G(k+1) = \mathbf{p}_b(k) - A\mathbf{s}, \tag{4.20}$$

where k corresponds to the actual iteration (generation), A and C represents two fixed coefficients, calculated based on Eqs. 4.12 and 4.13. $\mathbf{p}_b$ and $\mathbf{p}_G$ correspond to the positions of the best solution and another whale, respectively.

4.3.2.2 Bubble-Net Attacking Strategy

Under this strategy, the humpback whales employ two different mechanisms: shrinking (I) encircling and (II) spiraling. Both procedures contemplate the adjustment of positions over the best solution been so far. In WOA, it is supposed that both mechanisms keep an equal probability p_r to be performed. These mechanisms can be model as follows [R11]:

$$\mathbf{p}_G(k+1) = \begin{cases} \mathbf{p}_b(k) - A\mathbf{s} & \text{(I) if } p_r < 0.5 \\ \mathbf{s} \cdot \exp(bl) \cdot \cos(2\pi l) + \mathbf{p}_b(k) & \text{(II) otherwise} \end{cases}, \tag{4.21}$$

where b is a fixed coefficient that determines the logarithmic distribution l and p_r corresponds to two random numbers from the range [0, 1].

4.3.3 Crow Search Algorithm (CSA)

The Crow Search Algorithm (CSA) [17] is a well-known optimization method which is based on the interesting and complex abstraction process of crows when they are looking for food locations. In the CSA, each individual $\mathbf{p}_c$ ($\mathbf{p}_c \in \mathbf{P}(k)$) mimics the position of a crow in the group. Additionally, to each crow $\mathbf{p}_c$, there exists a memory structure $\mathbf{m}_c$ which holds the best solution along the evolutionary process.

In the CSA implementation, the position of each crow $\mathbf{p}_c$ is modified regarding two distinct behaviors: (I) Pursuit and (II) evasion.

(I) Pursuit. Under this procedure, a crow $\mathbf{p}_j$ pursues another crow $\mathbf{p}_i$ with the goal of finding its hidden food.

(II) Evasion. The crow $\mathbf{p}_i$ approximates to the position of crow $\mathbf{p}_j$. Consequently, the crow $\mathbf{p}_i$ intentionally chose a random trajectory to protect its food. This mechanism is computationally implemented using the assignment of a random position.

Each search individual $\mathbf{p}_i$ picks one of the aforementioned behaviors based on an awareness probability (AP). For that, a random number r_i is computed uniformly from range $[0, 1]$. Hence, when r_i is greater or equal than AP, the pursuit action (I) is executed; otherwise, the evasion behavior (II) is executed. This process is defined as:

$$\mathbf{p}_i(k+1) = \begin{cases} \mathbf{p}_i(k) + r_i \cdot fl_c\big(\mathbf{m}_j(k) - \mathbf{p}_c(k)\big) & \text{(I) if } r_i > AP \\ \text{random position} & \text{(II) otherwise} \end{cases}, \tag{4.22}$$

where fl_c corresponds the flight length adjustment of the crow $\mathbf{p}_c$ to the best position $\mathbf{m}_j$ of the crow $\mathbf{p}_i$.

4.3.4 Gravitational Search Algorithm (GSA)

The gravitational search algorithm (GSA) [22] is an abstraction based on the laws of gravitation. In GSA, the first stage the algorithm computes the total gravitational force of the system. It corresponds to the influence attraction force of the best solution over the remaining elements of the system, In the algorithm, K corresponds to the total number of individuals in the population defined as N_i. Also, each individual $\mathbf{p}_i(k)$ contains a force value $\mathbf{F}_i$ computed as:

$$\mathbf{F}_i(k) = \sum_{j \in K, j \neq i} r_2 G(k) \frac{M_j(k)M_i(k)}{R_{i,j}} \big(\mathbf{p}_j(k) - \mathbf{p}_i(k)\big), \tag{4.23}$$

$$G(k) = 100\exp\left(-20\frac{k}{Maxgen}\right), \tag{4.24}$$

where $i \in \{1, \ldots, N_i\}$; k represents the actual generation (iteration); r_2 represents a random value uniformly chosen from range $[0, 1]$; $R_{i,j}$ corresponds to the Euclidian distance among the individuals $\mathbf{p}_i(k)$ and $\mathbf{p}_j(k)$. $M_j(k)$ and $M_i(k)$ are calculated by:

$$M_i(k) = \frac{q_i(k)}{\sum_{v=1}^{N_i} q_v(k)} \tag{4.25}$$

$$q_i(k) = \frac{J(\mathbf{p}_i(k)) - J(\mathbf{a})}{J(\mathbf{b}) - J(\mathbf{a})}, \tag{4.26}$$

where $\mathbf{b}$ and $\mathbf{a}$ correspond to the best and the worst elements, respectively. Once the vector force is computed by each search agent, the acceleration of each individual $\mathbf{u}_i(k)$ is calculated, according to:

$$\mathbf{u}_i(k) = \frac{\mathbf{F}_i(k)}{M_i(k)} \tag{4.27}$$

Finally, the position of individual is adjusted as:

$$\mathbf{p}_i(k+1) = \mathbf{p}_i(k) + \mathbf{v}_i(k+1)$$

$$\mathbf{v}_i(k+1) = r_3\mathbf{v}_i(k) + \mathbf{u}_i(k), \tag{4.28}$$

where r_3 represents a uniform random number from the range [0, 1].

4.3.5 Cuckoo Search (CS) Method

The Cuckoo Search (CS) [18] algorithm is inspired by the fascinating behavior of cuckoo birds. In the CS operation, Levy flight random walk and biased random walk are used as the basis for its searching mechanism. Considering the Levy flight random walk, an adjustment is obtained by sampling a Levy distribution as:

$$Levy(\lambda) = \frac{u}{|v|^{1/\beta}} \tag{4.29}$$

where β corresponds to a parameter of the distribution which corresponds to a coefficient from the range [1, 2]. On the other hand, u and v follow a normal distribution strategy as follows:

$$u \sim N(0, \sigma_u^2), v \sim N(0, 1)$$

$$\sigma_u^2 = \left[\frac{\Gamma(1+\beta)\sin\left(\frac{\pi\beta}{2}\right)}{\Gamma\left(\frac{1+\beta}{2}\right)\beta 2^{(\beta-1)/2}} \right]^{1/\beta}, \tag{4.30}$$

where $\Gamma(\cdot)$ corresponds to the Gamma function. Hence, the new position of a search individual $\mathbf{p}_i(k+1)$ considering the Levy flight, random walk can be defined as:

$$\mathbf{p}_i(k+1) = \mathbf{p}_i(k) + step \oplus Levy(\lambda) \tag{4.31}$$

$$step = \alpha \oplus (\mathbf{p}_i(k) - \mathbf{b}) \tag{4.32}$$

where α corresponds to the step size and $\oplus$ represents the pairwise multiplication. Finally, the procedure of the biased random walk can be defined as:

$$\mathbf{p}_i(k+1) = \begin{cases} \mathbf{p}_i(k) + r_4(\mathbf{p}_m(k) - \mathbf{p}_n(k)) & r_5 > p_a \\ \mathbf{p}_i(k) & \text{otherwise} \end{cases} \tag{4.33}$$

where r_4 and r_5 represents uniform random values from the range $[0, 1]$ while p_a corresponds to the discovery probability. $\mathbf{p}_m(k)$ and $\mathbf{p}_n(k)$ represents two random individuals from the population at k iteration (generation) $\mathbf{P}(k)$.

4.4 Comparative Perspective of the Five Metaheuristic Methods

The five most popular evolutionary optimization algorithms presented in this chapter include: Cuckoo Search (CS), Grey Wolf Optimizer (GWO), Gravitational Search Algorithm (GSA), Crow Search Algorithm (CSA), and Whale Optimizer Algorithm (WOA) which are developed under distinct search mechanisms. In this section, a comparative analysis of these techniques is presented. The discussion evaluates the noticeable characteristics of each evolutionary method and how their features present an effect on their performance.

In the evolutionary computation community, it is widely accepted that a suitable balance among exploration and exploitation stages defines the potential of an evolutionary approach [34]. Exploration stage refers to the procedure of searching for new solutions into the search space. Additionally, exploitation stage considers the operation of tuning previously computed solutions to improve their fitness quality employing local search operators. The employment of only exploration decreases the accurateness of the solutions but increases the ability of the algorithm to acquire new profitable solutions [35]. On the contrary, the use of only exploitation operators allows raising the precision of existing solutions. Nonetheless, causes that the optimization process converges prematurely towards sub-optimal solutions [36]. The rate among exploration and exploitation used by certain evolutionary approach directly relies on the procedure of its search strategy.

For the study of the used evolutionary approaches, it has been considered the analysis of their search mechanisms as well as their computational implementations. The search mechanisms have been divided into two distinct parts: The selection process or attraction strategy and autonomy on the evolution process.

The selection mechanism refers to how each evolutionary approach incorporates solutions between the entire population to construct new solutions. Such chosen solutions are commonly employed as principal components to construct attraction operators to adjust the locations of the present population. There exist two distinct techniques [34]: greedy, random and complete. Greedy selection regards the benefit of the best individual or elements in terms of fitness values. Greedy selection guarantees that most fitting solutions stay intact through the optimization process. Under this, this technique facilitates the exploitation of the search space, rising fast convergence into profitable solutions. The algorithms GWO and WOA are evaluated within this classification. The WOA employs the best individual as a crucial individual in its search mechanism. In the GWO algorithm, the selection approach employs the three best entry elements. This strategy permits that not only select exploitation operators but also it helps by increasing the population diversity. Also, random selection regards the use of an individual randomly chosen within the population. The use of random elements facilitates the exploration of the search space. However, it affects the process of obtaining good accurateness. On the other hand, the CSA and CS algorithms consider the benefit of using a random solution as an important aspect of their search procedure. At last, the entire selection regards the usage of all individuals within the population to obtain newer solutions. This method employs the averaged location of the entire population to generate search direction trajectories. This strategy conducts the search into very low search scenarios where an additional number of iterations is required. On the other hand, the GSA method utilizes the whole selection strategy as in its search scheme.

Concerning the autonomy of the evolution procedure [37], two distinct procedures are denoted: Dependent and independent. Dependent procedures describe scenarios in which the number of generations is employed to adjust the search technique. The goal under these methods is to adjust the search method so that at the start of the optimization process, the exploration of the search space would be boosted while in the ending stage, the exploitation of the discovered solutions would be preferred [38]. Representative methods of this class are the well-known CSA and GSA. Dependent techniques offer several hardships since the outcomes of the search procedure are not believed in the procedure; it is more meaningful the number of generations. Independent algorithms do not assume that the stage of the optimization process guide their search processes. These methods use the knowledge made by the accepted outcomes to employ their search techniques [39]. In general words, independent approaches produce better results than dependent approaches. The algorithms CS, WOA and GWO are cataloged as independent methods.

Based on an implementation perspective, it is compared to the hardship that describes the benefit and operation of each evolutionary method. The methods are categorized regarding their implementation problems [R1] in low, medium and high. Such classes consider two aspects: the amount of tunning parameters and the extension of code required for its programming phase. The issue of parameter tuning relies on the number of configurable parameters [40]. It is essential to state that a proper configuration obtains a valid balance among exploitation and exploration operators therefore, competitive experimental results. Nowadays, there is no standard procedure that permits the suitable setting of an evolutionary approach. Researchers and the entire scientific community set the parameters of evolutionary methods by hand, guided just by the procedure of trial and error. For that, the GWO and WOA are the optimization algorithms that do not assume any parameter to be configured. The CS

Table 4.2 Comparison among search methodologies

Algorithm	Search mechanism		Implementation
	Selection or attraction scheme	Independence on the evolution process	
GWO	Greedy	Independent	Low
WOA	Greedy	Independent	Low
CSA	Random	Dependent	Medium
CS	Random	Independent	High
GSA	Complete	Dependent	High

and CSA consider 4 tuning parameters, while the GSA operates three components. Among the five methodologies analyzed in this investigation, some deliver a high level of simplicity so that they can be promptly translated into code with relative easiness [41]. In distinction to others are relatively complex due to the behaviors and rules considered in the model. According to these levels, the extension of required code to implement a specific evolutionary method can be employed as a standard to classify each approach as complex and simple. Thus, the CSA, WOA, and GWO deliver an easy level of programming while GSA and CS sustain a complex programming representation. Regarding to their implementation by code, the approaches evaluated in this research can be categorized as follows: GWO and WOA as low, CSA as medium and CS and GSA as high. Table 4.2 recaps all the comparative factors examined for all the tested algorithms.

4.5 Experimental Simulation Results

In this section, the study of the experimental simulations has been presented to compare the performance of the evolutionary methods considering the ellipse detection problem. The simulations evaluate the capability of each algorithm to identify elliptical shapes.

4.5.1 Performance Metrics

In the comparison, it is presented two of the most used performance metrics: the Multiple Error (ME) and the computational time (CT). The first metric measures the accuracy of the elliptical shape detection, while the second metric evaluates the computational complexity.

In real-world digital imaging system, ellipses are not considered as perfect shapes. For that, to evaluate the performance of the precision in the detection scheme, the estimated elliptical shape must be is compared considering ground-truth elliptical shapes which are obtained from the original image using a visual perspective. As a result, the parameters x_0^{true}, y_0^{true}, r_{max}^{true}, r_{min}^{true} and θ_{true} of the ground-truth elliptical shapes are defined based on the best-fitted shape that human eye can identify by employing a sketching program, for example: MSPaint, CorelDraw, etc. Therefore, if the parameters x_0^D, y_0^D, r_{max}^D, r_{min}^D and θ_D represent a detected elliptical shape, then the Error Score (Es) is computed as:

$$Es = c_1 \cdot CD + c_2 \cdot RM + c_3 \cdot AE$$
$$CD = \left| x_0^{true} - x_0^D \right| + \left| y_0^{true} - y_0^D \right|, RM = \left| r_{max}^{true} - r_{max}^D \right| + \left| r_{min}^{true} - r_{min}^D \right|,$$
$$AE = |\theta_{true} - \theta_D|, \tag{4.34}$$

In Eq. 4.34, CD represents the central point among the estimated elliptical shape and the ground-truth elliptical shape. RM corresponds to the radio tolerance while AE indicates the angle error among elliptical shapes. The coefficient c_1, c_2 and c_3 define the relative importance of each deviation CD, RM and AE in the final error Es. In the experimental simulations, the coefficients c_1, c_2 and c_3 are configured to 0.05, 0.1 and 0.2; respectively. Under this parameter settings, the angle error AE is strongly weighted respect to CD and RM. To prolong the Error Score (Es) value for multiple elliptical recognition, the averaged Es obtained by the position of multiple shapes is considered as the performance metric. This is called as the multiple error (ME) and it is calculated as:

$$ME = \frac{1}{NC} \sum_{q=1}^{NC} Es_q, \tag{4.35}$$

where NC represents the number of elliptical shapes stored in the edge map. Hence, Es_q corresponds to the Error Score acquired by the q elliptical shape from the original image. A lower ME value represents better exactness of the evolutionary approach.

In general, evolutionary optimization techniques are considered complex stochastic methods with different evolutionary operators and random branches. Hence, it results difficult to determine its performance. As a consequence, the computational time (CT) is measured to calculate the computational effort. Under this situation, CT represents the CPU time consumed by the execution of an evolutionary algorithm.

4.5.2 Experimental Comparison Study

In this section, the evolutionary methods considered in the study: Grey Wolf Optimizer (GWO) [20], the Whale Optimizer Algorithm (WOA) [21], the Crow Search Algorithm (CSA) [17], the Gravitational Search Algorithm (GSA) [22] and the Cuckoo Search (CS) [18] are numerically compared considering the performance metrics. The experimental simulations consider the capability of each methodology to identify overlapped or incomplete elliptical shapes including elliptical arcs segments. In the study, the population size N_i has been configured to 50 agents for each methodology. To compare the performance of the evolutionary methods, the configuration parameters have been configured, according to their reported original references. The implementation for each method has been obtained from their original sources.

The numerical results are divided into two different parts. In the first part (Sect. 4.5.2.1), the performance of the methodologies is evaluated considering the ellipse identification within real images. In the second part (Sect. 4.5.2.2), the robustness in the identification process of each algorithm in the presence of complex situations is studied.

4.5.2.1 Detection in Real Images

In the comparison study, all the evolutionary methodologies have been run over a fixed number of 1000 generations ($k = 1000$). This stop criterion has been selected to preserve compatibility with reported literature [7–13]. To minimize the random effect in the experimental results, each simulation is tested over 30 independent runs.

In the analysis, a representative group of 4 distinct images gathered from the literature have been employed. Figure 4.3 illustrates the complete group of images. In the figure, each image is tagged as D1.4 that refers to the image 4 from experiment one. In the analysis, the ground-truth elliptical shapes have been visually made from their original images.

The visual outcomes in the execution of all algorithms are shown in Fig. 4.4. Since each evolutionary method is executed 30 times over the same image, the recognition results are presented in terms of the average multiple error ($\overline{ME}$) and its standard deviation (SD) in Table 4.1. Regarding the recognition results from Fig. 4.4, it can be clearly shown that the GWO procedure delivers the best qualitative exactness followed by the WOA and the GSA whereas the CSA and the CS obtain the worst results.

Analyzing Table 4.3, it can be shown that considering the average multiple error ($\overline{ME}$) metric, the Grey Wolf Optimizer (GWO) algorithm outperforms better than the

| D1.1 | D1.2 | D1.3 | D1.4 |

Fig. 4.3 Set of real images for the experimental study

Fig. 4.4 Visual results

rest of methodologies. The Whale Optimizer Algorithm (WOA) keeps the second-best scores based on the $\overline{ME}$ performance metric. The Crow Search Algorithm (CSA) and the Cuckoo Search (CS) obtain the worst identification results. This fact highlights that WOA and GWO achieve a better balance among exploration stage and exploitation stage. Their search mechanisms permit discovering the best solution. Also, the remaining of the algorithms are attracted to local optima values. Regarding the standard deviation SD metric, it is clear CS holds the best consistency. The GWO achieves the second best SD value. Even when both algorithms GWO and CS present a low SD metric, their performance is distinct. In case of the CS algorithm, its search mechanism conducts the identification process towards single promissory local zone. For that, this sub-optimal solution is obtained each single run, producing

Table 4.3 Numerical results of the tested methods

		D1.1	D1.2	D1.3	D1.4
GWO	$\overline{ME}$	0.2750	0.1991	0.1051	0.1679
	SD	0.0220	0.0194	0.0067	0.0205
WOA	$\overline{ME}$	0.3245	0.2298	0.1185	0.1921
	SD	0.0304	0.0404	0.0098	0.0205
CSA	$\overline{ME}$	0.9272	0.9250	0.9250	0.7793
	SD	0.0591	0.0435	0.0421	0.0719
GSA	$\overline{ME}$	0.4913	0.4658	0.2509	0.2426
	SD	0.0536	0.0435	0.0421	0.0719
CS	$\overline{ME}$	0.8442	0.6915	0.9381	0.9991
	SD	0.0381	0.0536	0.0009	0.0001

a low SD value. Contrary to CS, the GWO technique achieves the best possible solution. However, due to its attraction operation, the solution is located towards the best solution. This relocation process causes a small dispersion from the optimal location. As a result, it can be concluded that the GWO maintains the best trade-off among consistency (SD) and accuracy ($\overline{ME}$).

Once recognized the GWO method is the best elliptical shape approach, statistical validation of the numerical results must be conducted. The non-parametric analysis known as the Wilcoxon test [42, 43] has been applied to validate the numerical results. The Wilcoxon rank-sum test allows assessing the performance distinctions among two different algorithms. The test is guided considering 5% (0.05) of significance level over 30 independent runs. Table 4.4 shows the p-values obtained by Wilcoxon test for a pair-wise comparison. For the test, four groups of comparisons are validated: GWO versus CSA, GWO versus WOA, GWO versus CS and GWO versus GSA. Regarding the Wilcoxon rank-sum test, it is considered as a null hypothesis that each comparison group does not show a significant statistical difference among them. Oppositely, the alternative hypothesis is accepted if both methodologies carry a considerable difference.

To visually demonstrate the experimental results of the Wilcoxon rank-sum test in Table 4.4, the symbols ▲, ▼, and ► are defined. ▲ corresponds to that the GWO approach produces statistically better results than its competitor. Also, ▼ indicates

Table 4.4 p-values obtained by the rank-sum test of Wilcoxon

	GWO versus WOA	GWO versus CSA	GWO versus GSA	GWO versus CS
D1.1	2.25E-04▲	8.76E-11▲	1.43E-09▲	4.84E-07▲
D1.2	6.67E-03▲	5.32E-11▲	3.97E-09▲	7.73E-07▲
D1.3	5.08E-01►	2.01E-11▲	6.28E-09▲	1.09E-07▲
D1.4	9.33E-02►	3.22E-11▲	9.43E-10▲	8.05E-08▲

that GWO performs worse than its competitor. Finally, ▶ indicates that Wilcoxon rank-sum test is not able to distinct among compared methodologies.

Regarding to the p-values results in Table 4.4, it is quite evident that the entire set of p-values in the groups GWO versus CS, GWO versus GSA and GWO versus CSA are less than the significance level which indicates that there exists sufficient evidence to revoke the null hypothesis. As conclusion, this fact suggests that GWO approach supports better performance than the rest of methodologies. The case of the images D1.3 and D1.4, it is quite evident that the GWO method exhibits similar results than WOA approach. This observation is determined by checking the column GWO versus WOA, where the p-values are higher than 0.05 (▶). In this scenario, there is no statistical difference considering the performance results among GWO and WOA. The comparison of the $\overline{ME}$ metric cannot determine the identification performance of the evolutionary method. For that, a further study considering the computational effort for each method has been employed. The main goal is to evaluate the velocity at which each methodology reaches the optimal solution. In Table 4.3, it is presented the average computational time $\overline{CT}$ of each evolutionary technique over 30 independent runs.

Regarding to the numerical results from Table 4.5, it can be shown that the CSA algorithm acquires the best $\overline{CT}$ value, followed by the GWO, WOA, GSA and CS. However, the CSA method obtains the worst identification based on the $\overline{ME}$ metric. Hence, the GWO obtains the best balance among computational effort and accurateness.

4.5.2.2 Detection in Complex Scenarios

In this study, it is presented the capability for each evolutionary method to locate promissory optimal solutions considering imperfect elliptical shapes. This experiment is relevant since image anomalies commonly appears in artificial vision tasks. In this chapter, the ellipse recognition formulated as an optimization problem. Then, each optimization method locates the best elliptical shape that better match according to the fitness function values for each possible solution [44, 45].

In the experimentations, the evolutionary approaches have been run with a fixed number of 1000 generations. Each simulation is evaluated for 30 runs to minimize the random presence. In the study, a set of 4 representative and different images is

Table 4.5 Average computational time in seconds

	D1.1	D1.2	D1.3	D1.4
GWO	37.42	41.9	65.67	67.21
WOA	40.53	38.2	74.14	70.12
CSA	10.98	21.07	30.54	40.87
GSA	73.05	51.14	87.25	92.47
CS	85.12	67.29	97.18	128.14

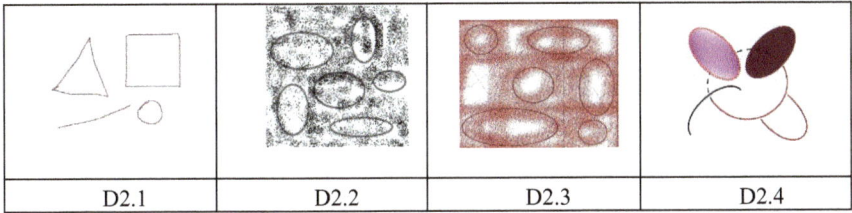

| D2.1 | D2.2 | D2.3 | D2.4 |

Fig. 4.5 Set of tested images for complex scenarios

collected from the literature. Figure 4.5 illustrates the entire group of test images. In the figure, each image is tagged as D2.4 that means that the image 4 from experiment two ad so on. In the analysis, the ground truth outcomes for original images have been visually obtained.

The visual outcomes for all the algorithms are presented in Fig. 4.6. Since every evolutionary method is evaluated 30 runs, the identification results are presented considering the $\overline{ME}$ metric and its SD value from Table 4.6. From the recognition results presented in Fig. 4.6, it is clear that GWO acquires the best qualitative accurateness followed by the WOA and the GSA methodologies, while the CSA and the CS obtain the worst recognitions.

Considering the numerical results exposed in Table 4.6, the remarkable performance of GWO over CSA, GSA, and CS is again demonstrated in terms $\overline{ME}$ value. Nevertheless, the GWO is just barely better than WOA procedure. Similar to the case of real images, the CS method obtains solutions with the smallest level of next by the GWO algorithm. This robustness is correct for all images D2.1, D2.2, D2.3 and D2.4. The outstanding outcomes of GWO are achieved based on two different factors. The fist one of them is based on the attraction procedure. This strategy allows elitism and diversity at the same time.

As conclusion, it is conceivable to locate the best elliptical shape even in complex scenarios. This effect can be confirmed by the error metric $\overline{ME}$, which suggests that GWO achieves the best accurateness between the remaining approaches. Additionally, the second important factor is based on its implementation features. GWO does not require lots of code lines for its implementation and it does not require tunned parameters. For that, it is conceivable to adopt the usage of the GWO algorithm to multiple applications.

The statistical results for the Wilcoxon rank-sum test for the recognition of elliptical shapes considering complex scenarios are exposed in Table 4.7. Considering the outcomes of Table 4.7, it can be shown that the p-values in the comparisons of the groups GWO versus CS, GWO versus GSA, and GWO versus CSA. are less than the significance level, which means that GWO outperforms the remaining methods. In the case of the group comparison of GWO versus WOA, the GWO algorithm exhibits better performance for image D2.4, however, in images D2.1, D2.2, and D1.3 the WOA delivers equivalent performance as the GWO procedure.

To incorporate computational effort analysis in complex scenarios, Table 4.8 illustrates the $\overline{CT}$ of each evolutionary technique over 30 runs for each image. As it can

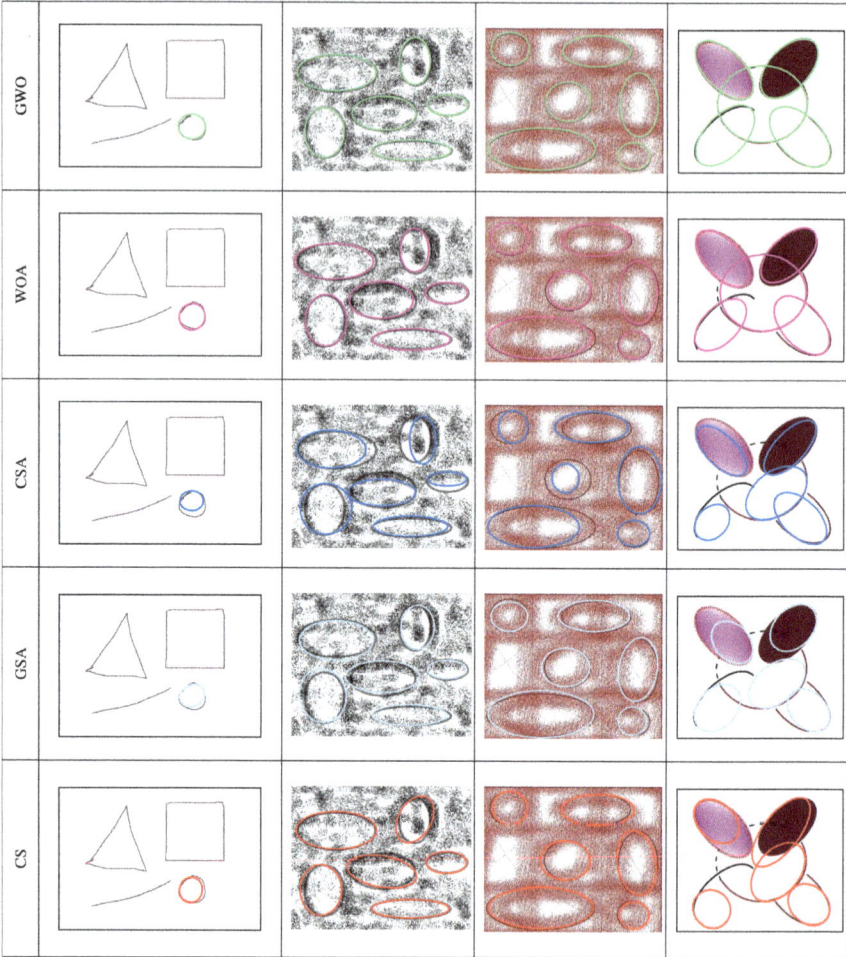

Fig. 4.6 Visual results in complex scenarios

be shown, the CSA approach acquire the best scores, followed by the GWO, WOA, GSA and CS. For that, the GWO delivers the best proportion among computational cost and accurateness.

4.6 Conclusions

In this chapter, a comparative study of the elliptical shape recognition problem has been presented. In the analysis, the Cuckoo Search (CS), Crow Search Algorithm (CSA), Whale Optimizer Algorithm (WOA), Grey Wolf Optimizer (GWO), and

Table 4.6 Identification results

		D2.1	D2.2	D2.3	D2.4
GWO	$\overline{ME}$	0.5388	0.09859	0.2974	0.3454
	SD	0.0215	0.0368	0.7253	0.0193
WOA	$\overline{ME}$	0.5494	0.2298	0.1185	0.1921
	SD	0.0335	0.0385	1.2178	0.0208
CSA	$\overline{ME}$	0.9225	0.8180	0.9790	0.9140
	SD	0.0514	0.0984	1.5961	0.0507
GSA	$\overline{ME}$	0.7072	0.1779	0.6915	0.4911
	SD	0.0361	0.0488	1.3925	0.0570
CS	$\overline{ME}$	0.8012	0.2813	0.4494	0.6330
	SD	0.0157	0.0225	0.4781	0.0313

Table 4.7 p-values obtained by the rank-sum test of Wilcoxon in complex scenarios

	GWO versus WOA	GWO versus CSA	GWO versus GSA	GWO versus CS
D2.1	8.24E−02▶	3.02E−11▲	5.64E−07▲	3.21E−08▲
D2.2	6.73E−01▶	1.41E−11▲	2.88E−10▲	4.46E−08▲
D2.3	2.40E−01▶	4.98E−11▲	2.97E−10▲	5.10E−08▲
D2.4	7.95E−03▲	1.54E−11▲	7.18E−11▲	6.38E−09▲

Table 4.8 Average computational time in seconds considering complex scenarios

	D1.1	D1.2	D1.3	D1.4
GWO	29.49	6.87	7.39	21.95
WOA	32.41	8.51	8.24	24.37
CSA	9.41	2.74	1.76	10.03
GSA	37.21	12.87	10.52	29.97
CS	35.23	11.43	10.87	27.08

Gravitational Search Algorithm (GSA) have been compared. These evolutionary methods have been commonly employed in multiple research fields. For that, the main goal is to compare their performance considering the complexity of elliptical shape recognition.

To supply a fair performance analysis of the ellipse recognition problem, it is considered as a complex optimization task- For that, the comparison study in this chapter is separated into two different parts. In the first section, the performance of the evolutionary methods is measured evaluated over real-world images. The second part, the robustness in the recognition of approximated and original elliptical shapes is presented. In this examination, multiple performance metrics have been computed.

Experimental results revealed that Grey Wolf Optimizer (GWO) is capable of providing the best performance in terms of the performance indexes. Additionally, this conclusion has been confirmed according to a non-parametric approach.

References

1. Bai X, Yang X, Latecki LJ (2008) Detection and recognition of contour parts based on shape similarity. Pattern Recogn 41(7):2189–2199
2. Schindler K, Suter D (2008) Object detection by global contour shape. Pattern Recogn 41(12):3736–3748
3. Lu W, Tan J (2008) Detection of incomplete ellipse in images with strong noise by iterative randomized Hough transform (IRHT). Pattern Recogn 41:1268–1279
4. Ayala-Ramirez V, Garcia-Capulin CH, Perez-Garcia A, Sanchez Yanez RE (2006) Circle detection on images using genetic algorithms. Pattern Recogn Lett 27:652–657
5. Lutton E, Martinez P (1994) A genetic algorithm for the detection of 2D geometric primitives in images. In: Proceedings of the 12th international conference on pattern recognition, Jerusalem, Israel 1, vol 14, pp 526–528
6. Jie Y, Nawwaf K, Peter G (2005) A multi-population genetic algorithm for robust and fast ellipse detection. Pattern Anal Appl 8:149–162
7. Cheng HD, Yanhui G, Yingtao Z (2009) A novel Hough transform based on eliminating particle swarm optimization and its applications. Pattern Recogn 42(9):1959–1969
8. Das S, Dasgupta S, Biswas A, Abraham A (2008) Automatic circle detection on images with annealed differential evolution. In: Eighth international conference on hybrid intelligent systems, 2008, HIS '08
9. Cuevas E, Osuna-Enciso V, Oliva D (2015) Circle detection on images based on the Clonal Selection Algorithm (CSA). Imaging Sci J 63(1):34–44
10. Fourie J (2017) Robust circle detection using harmony search. J Optim 2017, Article ID 9710719, 11
11. Cuevas E, Sención-Echauri F, Zaldivar D, Pérez-Cisneros M (2012) Multi-circle detection on images using artificial bee colony (ABC) optimization. Soft Comput 16(2):281–296
12. Cuevas E, González M, Zaldívar D, Pérez-Cisneros M (2014) Multi-ellipses detection on images inspired by collective animal behavior. Neural Comput Appl 24(5):1019–1033
13. Cuevas E, González M (2013) Multi-circle detection on images inspired by collective animal behavior. Appl Intell 39(1):101–120
14. Piotrowski A, Napiorkowski J (2018) Some metaheuristics should be simplified. Inf Sci 427:32–62
15. Camargo MP, Rueda JL, Erlich I, Añó O (2014) Comparison of emerging metaheuristic algorithms for optimal hydrothermal system operation. Swarm Evolut Comput 18:83–96
16. Dubey HM, Pandit M, Panigrahi BK (2018) An overview and comparative analysis of recent bio-inspired optimization techniques for wind integrated multi-objective power dispatch. Swarm Evolut Comput 38:12–34
17. Askarzadeh A (2016) A novel metaheuristic method for solving constrained engineering optimization problems: crow search algorithm. Comput Struct 169:1–12
18. Yang XS, Deb S (2009) Cuckoo search via Lévy flights. In: 2009 World congress on nature and biologically inspired computing, NABIC 2009—proceedings, pp 210–214
19. Mirjalili S (2015) Moth-flame optimization algorithm: a novel nature-inspired heuristic paradigm. Knowl Based Syst 89:228–249
20. Mirjalili S, Mirjalili SM, Lewis A (2014) Grey wolf optimizer. Adv Eng Softw 69:46–61
21. Mirjalili S, Lewis A (2016) The whale optimization algorithm. Adv Eng Softw 95:51–67
22. Rashedi E, Nezamabadi-pour H, Saryazdi S (2009) GSA: a gravitational search algorithm. Inf Sci (NY) 179(13):2232–2248

23. Mirjalili S (2016) SCA: a sine cosine algorithm for solving optimization problems. Knowl Based Syst 96:120–133
24. Canayaz M, Karci A (2016) Cricket behaviour-based evolutionary computation technique in solving engineering optimization problems. Appl Intell 44:362–376
25. Mirjalili S (2016) Dragonfly algorithm: a new meta-heuristic optimization technique for solving single-objective, discrete, and multi-objective problems. Neural Comput Appl 27(4):1053–1073
26. Zhou Y, Su K, Shao L (2018) A new chaotic hybrid cognitive optimization algorithm. Cogn Syst Res 52:537–542
27. Mortazavi A, Toğan V, Nuhoğlu A (2018) Interactive search algorithm: a new hybrid metaheuristic optimization algorithm. Eng Appl Artif Intell 71:275–292
28. Shayanfar H, Gharehchopogh FS (2018) Farmland fertility: a new metaheuristic algorithm for solving continuous optimization problems. Appl Soft Comput 71:728–746
29. Beşkirli M, Koç İ, Haklı H, Kodaz H (2018) A new optimization algorithm for solving wind turbine placement problem: binary artificial algae algorithm. Renew Energy 121:301–308
30. Tabari A, Ahmad A (2017) A new optimization method: electro-search algorithm. Comput Chem Eng 103:1–11
31. Al-Betar MA, Awadallah MA (2018) Island bat algorithm for optimization. Expert Syst Appl 107:126–145
32. Bresenham JE (1987) A linear algorithm for incremental digital display of circular arcs. Commun ACM 20:100–106
33. Caraveo C, Valdez F, Castillo O (2016) Optimization of fuzzy controller design using a new bee colony algorithm with fuzzy dynamic parameter adaptation. Appl Soft Comput 43:131–142
34. Fausto F, Reyna-Orta A, Cuevas E, Andrade ÁG, Pérez-Cisneros M (2020) From ants to whales: metaheuristics for all tastes. Artif Intell Rev 53:753–810
35. Cuevas E, Reyna-Orta A, Díaz-Cortes M-A (2018) A multimodal optimization algorithm inspired by the states of matter. Neural Process Lett 48:517–556
36. Tan KC, Chiam SC, Mamun AA, Goh CK (2009) Balancing exploration and exploitation with adaptive variation for evolutionary multi-objective optimization. Eur J Oper Res 197:701–713
37. Du H, Wang Z, Zhan WEI (2018) Elitism and distance strategy for selection of evolutionary algorithms. IEEE Access 6:4453144541
38. Morales-Castañeda B, Zaldívar D, Cuevas E, Fausto F, Rodríguez A (2020) A better balance in metaheuristic algorithms: does it exist? Swarm Evol Comput 54:100671
39. Črepinšek M, Liu S-H, Mernik M (2013) Exploration and exploitation in evolutionary algorithms. ACM Comput Surv 45(3):1–33
40. Yang X-S (2018) Swarm-based metaheuristic algorithms and no-free-lunch theorems. Intech Open 2:64
41. Piotrowski AP, Napiorkowski JJ (2018) Some metaheuristics should be simplified. Inf Sci (NY) 427:3262
42. Wilcoxon F (1945) Individual comparisons by ranking methods. Biometrics 1:80–83
43. Garcia S, Molina D, Lozano M, Herrera F (2008) A study on the use of non-parametric tests for analyzing the evolutionary algorithms' behavior: a case study on the CEC'2005 Special session on real parameter optimization. J Heurist. https://doi.org/10.1007/s10732-008-9080-4
44. Cuevas E, González A, Fausto F, Zaldívar D, Pérez-Cisneros M (2015) Multithreshold segmentation by using an algorithm based on the behavior of locust swarms. Math Probl Eng 2015:805357
45. Cuevas E, Zaldivar D, Pérez-Cisneros M (2011) Seeking multi-thresholds for image segmentation with learning automata. Mach Vis Appl 22(5):805–818

Chapter 5
IIR System Identification Using Several Optimization Techniques: A Review Analysis

System identification is a difficult optimization problem, especially those that use the infinite impulse response (IIR) models which are preferred over their equivalent FIR (finite impulse response) models since they represent more accurate real-world applications. Nevertheless, IIR models tend to generate multimodal error surfaces which are significantly difficult to optimize. Evolutionary computation (EC) algorithms are employed to determine the best solution to complex problems. EC's are designed to satisfy the requirements of particular problems because no single optimization technique is able to solve all problems efficiently. Therefore, when a new optimization technique is proposed, its relative effectiveness must be properly evaluated. Some comparative studies between EC's have been described in the state-of-art. Nevertheless, they have some limitations: their conclusions are based on the performance over a set of synthetic functions without considering the application context or including new optimization techniques. In this chapter, a comparison of various optimization techniques employed for IIR model identification is presented. In the comparison, special attention is paid to the popular algorithms such as Cuckoo Search and Flower Pollination Algorithm. Results over different models are performed and statistically verified.

5.1 Introduction

System identification is an optimization problem that recently attracted attention in science and engineering. The system identification task is important in the fields of control systems [1], signal processing [2], communication [3], and image processing [4]. In the system identification process, an optimization technique attempts to iteratively estimate the model parameters to produce an optimal model for an unknown plant by the minimization of some cost function. The model generated or solution is achieved when such cost function is effectively optimized. The accuracy of the determined model depends on the adaptive model structure, the optimization technique, and the features of the input–output data [5].

Systems can be modeled by infinite impulse response (IIR) models because they emulate physical plants more precisely than the FIR (finite impulse response) models [6]. Moreover, IIR models are typically able to meet the performance specifications employing some model parameters. Nevertheless, IIR structures produce multimodal error surfaces which are significantly challenging to minimize [7]. Therefore, to identify IIR models, a robust global optimization technique is essential to optimize the multimodal error function. The least mean square (LMS) method and its modifications [8] have been widely employed as an optimization technique for IIR model identification. The acceptance of gradient-based optimization techniques is due to the simple implementation and low complexity. Nevertheless, the error surface for the IIR model is frequently multi-modal regarding the filter coefficients. This may result in leading classical gradient-descent approaches into local minimal [9].

The challenges associated with the gradient-based optimization approaches for solving many problems have contributed to the development of different solutions. Evolutionary computation (EC) algorithms as the particle swarm optimization (PSO) [10], artificial bee colony (ABC) [11], electromagnetism-like method (EM) [12], Cuckoo Search (CS) [13] and Flower Pollination Algorithm (FPA) [14] have been widely used due to their potential as global optimization techniques in different applications. Inspired optimization techniques are based on the evolution process and survival of the most beneficial in the natural world, EC's are search techniques that are different from traditional methods. Those techniques are based on a collective learning process considering a population of candidate solutions. This population is usually randomly initialized, where each iteration evolves towards better and better solution regions using stochastics processes where many operators are used to each element of the population. EC's have been applied to many optimization problems and have been demonstrated to be accurate for solving any particular problems, including multimodal optimization, dynamic optimization, just to mention a few [15–17].

As an alternative to gradient-based methods, the IIR modeling problem has also been handled through EC's techniques. Usually, they have shown to produce better results than those based on gradient techniques regarding accuracy and robustness [9]. EC's have provided some robust IIR identification systems by employing different EC techniques such as PSO [18], ABC [19], EM [20], and CS [21], which are reported below. EC's are frequently designed to satisfy the requirements of particular problems due to no single optimization technique can solve all problems [22]. Hence, when a new algorithm is proposed, its efficiency must be properly evaluated. Several efforts [23–25] have also been dedicated to comparing EC's to each other. Usually, the comparative studies are based on synthetic numerical benchmark test functions for verifying if one technique outperforms others overlooking any statistical test. Nevertheless, some comparative studies of EC's regarding the application context are reported in the literature. Consequently, it is very important to compare the performance and accuracy of EC's techniques from an application point of view.

This chapter presents a review analysis using the comparison of several evolutionary computation techniques for IIR model identification. In this study, special attention is considered to popular developed optimization techniques as the Cuckoo Search (CS) and the Flower Pollination Algorithm (FPA), including also the most used approaches as the Particle Swarm Optimization (PSO), the Artificial Bee Colony (ABC), and optimization and the Electromagnetism-Like Optimization (EM). The experimental results over different models with diverse ranges of complexity are presented and statistically validated.

5.2 Evolutionary Computation (EC) Algorithms

In real-world applications, several identification problems can be considered blackbox tasks. Usually, some data is available concerning an optimization problem itself except the knowledge emerging from function evaluations. In some cases, any information is known about the features of the fitness function as in unimodal or multimodal functions. On the other hand, EC's are employed to determine solutions to complex problems since they can easily adapt to black-box problems and extremely multimodal functions. EC's are based on a cooperative learning method into a population of possible solutions. The population in an EC is regularly randomly initialized while each generation evolves towards more reliable solution regions through stochastics processes with different operators being used to each solution. EC's are applied to several optimization problems guaranteeing an efficient solution for some particular problems, including multimodal tasks, dynamic optimization, multi-objective problems, just to mention a few [15–17].

Hence, EC's have become popular tools for solving different difficult problems. This subsection gives a short description of the five evolutionary computation techniques used in the study: Swarm Optimization (PSO), Artificial Bee Colony (ABC) Optimization, Electromagnetism-Like Optimization (EM), Cuckoo Search (CS), and Flower Pollination Algorithm (FPA).

5.2.1 Particle Swarm Optimization (PSO)

The PSO was proposed by Kennedy and Eberhart [10]. This technique is a population-based optimization algorithm inspired by the social behavior of bird flocking or fish schooling. The PSO searches for the best solution using a swarm formed by possible solutions which are called particles. From the computational point of view, a group $\mathbf{P}^k$ ($\{\mathbf{p}_1^k, \mathbf{p}_2^k, \ldots, \mathbf{p}_N^k\}$) of N elements (individuals) evolves from the initial point ($k = 0$) to a total number of generations ($k = gen$). Each element $\mathbf{p}_i^k$ ($i \in [1, \ldots, N]$) describes a d-dimensional vector $\{p_{i,1}^k, p_{i,2}^k, \ldots, p_{i,d}^k\}$ in which every dimension corresponds to the problem variables. The quality of every solution $\mathbf{p}_i^k$ is estimated by employing an objective function $f\left(\mathbf{p}_i^k\right)$ which the result describes the fitness value of $\mathbf{p}_i^k$. Through the evolving process, the best position $\mathbf{g}$ $(g_1, g_2, \ldots g_d)$ observed so far is saved as the best position $\mathbf{p}_i^*$ $(p_{i,1}^*, p_{i,2}^*, \ldots, p_{i,d}^*)$ of each particle. Those positions are calculated by considering the minimization of the problem as follows:

$$\mathbf{g} = \underset{i \in \{1, 2, \ldots, N\}, a \in \{1, 2, \ldots, k\}}{\arg \min} (f(\mathbf{p}_i^a)) \quad \mathbf{p}_i^* = \underset{a \in \{1, 2, \ldots, k\}}{\arg \min} (f(\mathbf{p}_i^a)). \tag{5.1}$$

In this study, the variant of PSO proposed by Lin et al. in [26] is implemented. Under such circumstances, the new position $\mathbf{p}_i^{k+1}$ of the population $\mathbf{p}_i^k$ is estimated using the following equations:

$$v_{i,j}^{k+1} = w \cdot v_{i,j}^k + c_1 \cdot r_1 \cdot (p_{i,j}^* - p_{i,j}^k) + c_2 \cdot r_2 \cdot (g_j - p_{i,j}^k);$$
$$p_{i,j}^{k+1} = p_{i,j}^k + v_{i,j}^{k+1}; \tag{5.2}$$

w is the inertia weight that guides the influence of the current velocity on the new velocity $(i \in [1, \ldots, N], j \in [1, \ldots, d])$. c_1 and c_2 are positive coefficients that control the action of each element towards the positions $\mathbf{g}$ and $\mathbf{p}_i^*$, respectively. r_1 and r_2 are uniformly distributed random numbers between the interval $[0, 1]$.

5.2.2   The Artificial Bee Colony (ABC)

The ABC technique was introduced by Karaboga [19], which is an EC motivated by the intelligent behavior of honeybees. In the ABC procedure, a group $\mathbf{l}^k$ $(\{\mathbf{l}_1^k, \mathbf{l}_2^k, \ldots, \mathbf{l}_N^k\})$ of N feasible food locations (individuals) is evolved from the initial point $(k = 0)$ to a total number of generations $(k = gen)$. Every food location $\mathbf{l}_i^k$ $(i \in [1, \ldots, N])$ describes a d-dimensional vector $\{l_{i,1}^k, l_{i,2}^k, \ldots, l_{i,d}^k\}$ where each dimension is a decision variable of the problem to be treated. After the initialization process, an objective function estimates each food location to assess whether it represents a good solution (nectar-amount) or not. Conducted by the quality of the position $\mathbf{l}_i^k$ is evolved by diverse ABC operations (honeybee types). The main operator produces a new food source for each food location $\mathbf{l}_i^k$ in the neighborhood of its actual position as follows:

$$\mathbf{t}_i = \mathbf{l}_i^k + \phi(\mathbf{l}_i^k - \mathbf{l}_r^k), i, r \in (1, 2, \ldots, N) \tag{5.3}$$

where $\mathbf{l}_r^k$ is a random food location which satisfies the condition $r \neq i$. The factor ϕ is a random number within the range $[-1, 1]$. When a new solution $\mathbf{t}_i$ is produced, a fitness value describing the profitability correlated with a particular solution $fit(\mathbf{l}_i^k)$ is computed. The fitness value for a minimization task can be attributed to a candidate solution $\mathbf{l}_i^k$ by the next expression:

$$fit(\mathbf{l}_i^k) = \begin{cases} \frac{1}{1+f(\mathbf{l}_i^k)} & \text{if } f(\mathbf{l}_i^k) \geq 0 \\ 1 + abs(f(\mathbf{l}_i^k)) & \text{if } f(\mathbf{l}_i^k) < 0 \end{cases} \tag{5.4}$$

where $f(\cdot)$ expresses the objective function to be optimized. When the fitness values are determined, a selection process is employed between $\mathbf{t}_i$ and $\mathbf{l}_i^k$. If $fit(\mathbf{t}_i)$ is better than $fit(\mathbf{l}_i^k)$, then the element is replaced by $\mathbf{l}_i^k$ is replaced by $\mathbf{t}_i$; otherwise, $\mathbf{l}_i^k$ remains.

5.2.3 The Electromagnetism-Like (EM) Technique

The EM technique is proposed by İlker-Birbil and Shu-Cherng [12]. This algorithm is a population-based method inspired by the electro-magnetism phenomenon In the EM procedure, the individuals emulate charged elements that interact with each other based on the laws of repulsion and attraction of the electro-magnetism theorem. The method employs N, d-dimensional objects $\mathbf{x}_i^k$, $i = 1, 2, \ldots, N$ where each object is a $\mathbf{x}_i^k$ is a d-dimensional vector containing the parameter values of the problem ($\mathbf{x}_i^k = \{x_{i,1}^k, \ldots, x_{i,d}^k\}$) whereas k indicates the generation number. The initial population $\mathbf{X}^k = \{\mathbf{x}_1^k, \mathbf{x}_2^k, \ldots, \mathbf{x}_N^k\}$ is selected from randomly distributed samples of the search space. We express the population set at the k-th iteration by $\mathbf{X}^k$, because members of $\mathbf{X}^k$ due to the elements of change is presented as k. After the initialization of $\mathbf{X}^0$, EM remains its evolving process until a stopping criterion is met. The evolution process of EM consists of three main steps. The first step moves each element $\mathbf{X}^k$ into different locations by applying the attraction–repulsion mechanism of the electromagnetism theory. The second step moves the elements by the electromagnetism system, additionally positioned locally for a local search procedure. Finally, in the third step, in order to generate the new population $\mathbf{X}^{k+1}$, a greedy selection process selects the best points between those produced by the local search procedure and the originals. Attraction–repulsion mechanisms and the local search in EM are responsible for handling the members $\mathbf{x}_i^k$ of $\mathbf{X}^k$ to the closeness of the optimum.

5.2.4 Cuckoo Search (CS) Technique

The CS method is one of the most popular nature-inspired techniques which has been introduced by Yang and Deb [21]. The CS algorithm is based on the family dependency of some cuckoo species. Moreover, this technique is improved by the method called Lévy flights [27], rather than by simple random walks. From the computational implementation, the CS generates a population $\mathbf{E}^k$ ($\{\mathbf{e}_1^k, \mathbf{e}_2^k, \ldots, \mathbf{e}_N^k\}$) of N eggs (elements) that evolves from the initial point ($k = 0$) to a total number of generations ($k = 2 \cdot gen$). Each egg $\mathbf{e}_i^k$ ($i \in [1, \ldots, N]$) describes a d-dimensional vector $\{e_{i,1}^k, e_{i,2}^k, \ldots, e_{i,d}^k\}$ where each dimension is a decision variable of the problem. The quality of each solution $\mathbf{e}_i^k$ (candidate solution) is evaluated by using an objective function $f\left(\mathbf{e}_i^k\right)$ whose result describes the fitness value of $\mathbf{e}_i^k$. Three main operators describe the optimization process of CS: (A) Lévy flight, (B) the Replace of Nests operator for creating new solutions, and (C) the selection procedure.

(A) The Lévy flight

$$s_i = \frac{\mathbf{u}}{|\mathbf{v}|^{1/\beta}}, \tag{5.5}$$

where $\mathbf{u}$ ($\{u_1, \ldots, u_d\}$) and $\mathbf{v}$ ($\{v_1, \ldots, v_d\}$) are n-dimensional vectors and $\beta = 3/2$. Every element of $\mathbf{u}$ and $\mathbf{v}$ are determined by the following distributions:

$$u \sim N(0, \sigma_u^2), \qquad v \sim N(0, \sigma_v^2),$$
$$\sigma_u = \left(\frac{\Gamma(1+\beta)\cdot\sin(\pi\cdot\beta/2)}{\Gamma((1+\beta)/2)\cdot\beta\cdot 2^{(\beta-1)/2}}\right)^{1/\beta}, \qquad \sigma_v = 1, \tag{5.6}$$

where $\Gamma(\cdot)$ denotes the Gamma distribution. After $\mathbf{s}_i$ has been determined, the essential change of position $\mathbf{c}_i$ is calculated as follows:

$$\mathbf{c}_i = 0.01 \cdot \mathbf{s}_i \oplus (\mathbf{e}_i^k - \mathbf{e}^{best}), \tag{5.7}$$

where $\oplus$ expresses the entry-wise multiplications, while $\mathbf{e}^{best}$ is the best solution (egg) observed in terms of fitness value. Lastly, the new solution is $\mathbf{e}_i^{k+1}$ is computed by employing:

$$\mathbf{e}_i^{k+1} = \mathbf{e}_i^k + \mathbf{c}_i \tag{5.8}$$

(B) Replace some nests by constructing new solutions

This operator generates a set of elements that are probabilistically chosen and replaced with a new one. Each element $\mathbf{e}_i^k$ ($i \in [1, \ldots, N]$) can be picked with a probability $p_a \in [0, 1]$. To perform this action, a random number r_1 is produced within the range $[0, 1]$. If r_1 is lower than p_a, the element $\mathbf{e}_i^k$ is chosen and adjusted according to Eq. 5.5; otherwise $\mathbf{e}_i^k$ remains with no change. This operation can be described by the following model:

$$\mathbf{e}_i^{k+1} = \begin{cases} \mathbf{e}_i^k + rand \cdot (\mathbf{e}_j^k - \mathbf{e}_h^k) & \text{with probability } p_a \\ \mathbf{e}_i^k & \text{with probability } (1 - p_a) \end{cases} \tag{5.9}$$

where $rand$ is a random number while j and h are random integers from 1 to N.

(C) Elitist Selection

After generating $\mathbf{e}_i^{k+1}$ by using the operator A or B, it must be compared with its latest value. $\mathbf{e}_i^k$. If the quality of $\mathbf{e}_i^{k+1}$ is better than $\mathbf{e}_i^k$, then $\mathbf{e}_i^{k+1}$ is taken as the final solution; otherwise, $\mathbf{e}_i^k$ remains. This method can be resumed as follows:

$$\mathbf{e}_i^{k+1} = \begin{cases} \mathbf{e}_i^{k+1} & \text{if } f(\mathbf{e}_i^{k+1}) < f(\mathbf{e}_i^k) \\ \mathbf{e}_i^k & \text{otherwise} \end{cases} \tag{5.10}$$

The selection strategy indicates that just high-quality eggs (best solutions near to the global optimal) which are the most similar to the host bird's eggs have the chance to improve (next generation) and grow mature cuckoos.

5.2.5 Flower Pollination Algorithm (FPA)

The FPA technique was proposed by Yang [14], which is an EC motivated by the pollination process of some flowers. In the FPA process, elements emulate a set of flowers or pollen gametes that behave based on biological laws of pollination. From the computational point of view, the FPA evolves a population $\mathbf{F}^k$ ($\{\mathbf{f}_1^k, \mathbf{f}_2^k, \ldots, \mathbf{f}_N^k\}$) flower positions from the initial point $(k = 0)$ to a total number of generations $(k = gen)$. Each element $\mathbf{f}_i^k$ ($i \in [1, \ldots, N]$) describes a d-dimensional vector $\{f_{i,1}^k, f_{i,2}^k, \ldots, f_{i,d}^k\}$, in the FPA a new population $\mathbf{F}^{k+1}$ is generated by taking into account two operators: local and global pollination. A global pollination factor p is correlated with these operators. To determine which operator should be employed to each current element position $\mathbf{f}_i^k$, a random number r_p is produced within the range [0, 1]. If r_p is lower than p, the global pollination operator is employed to $\mathbf{f}_i^k$. Otherwise, the local pollination operator is contemplated.

Global pollination operation

Under this operation, the first position $\mathbf{f}_i^k$ is replaced with a new position $\mathbf{f}_i^{k+1}$ according to the next model:

$$\mathbf{f}_i^{k+1} = \mathbf{f}_i^k + s_i \cdot (\mathbf{f}_i^k - \mathbf{g}) \tag{5.11}$$

where $\mathbf{g}$ is the global best observed so far while s_i controls the length of the movement. The s_i value is produced by a symmetric Lévy distribution considering Eqs. 5.5–5.6.

Local pollination operation

In the local pollination operation, the actual position $\mathbf{f}_i^k$ is modified to a new position $\mathbf{f}_i^{k+1}$ as below:

$$\mathbf{f}_i^{k+1} = \mathbf{f}_i^k + \varepsilon \cdot (\mathbf{f}_j^k - \mathbf{f}_h^k); i, j, h \in (1, 2, \ldots, N), \tag{5.12}$$

where $\mathbf{f}_j^k$ and $\mathbf{f}_h^k$ two random flower positions, which satisfies the condition $j \neq h \neq i$. The factor ε is a random number between $[-1, 1]$.

5.3   Formulation of IIR Model Identification

The system identification process is a mathematical description of an unknown system by employing only input–output information. In the system identification method, an optimization technique attempts to iteratively estimate the adaptive parameters model to get an optimal system model for the unknown plant based on minimizing the error between the output of the candidate system and the original output of the plant.

The use of infinite impulse response (IIR) models for system identification is favoured over their equivalent FIR (finite impulse response) models since they can produce more accurate responses of physical plants for real-world applications [6].

Moreover, IIR models are typically able to meet better performance specifications employing fewer design parameters. Figure 5.1 shows an IIR identification system of any arbitrary model.

An IIR model can be expressed by the following transfer function:

$$\frac{Y(z)}{X(z)} = \frac{b_0 + b_1 z^{-1} + b_2 z^{-2} + \cdots + b_m z^{-m}}{1 + a_1 z^{-1} + a_2 z^{-2} + \cdots + a_n z^{-n}} \qquad (5.13)$$

where m and n are the numerator and denominator coefficients of the transfer function respectively, a_i and b_j are the pole and zero parameters of the IIR system ($i \in [1, \ldots, n]$, $j \in [1, \ldots, m]$). Equation 5.13 can be expressed as difference equation as below:

$$y(t) = \sum_{i=1}^{n} a_i \cdot y(t-i) + \sum_{j=0}^{m} b_j \cdot x(t-j) \qquad (5.14)$$

where $u(t)$ and $y(t)$ denote the t-th input and output of the system, respectively. Consequently, a set of unknown parameters that models the IIR system is described by $\theta = \{a_1, \ldots, a_n, b_0, \ldots, b_m\}$. Taking into account that the number of parameters of θ is $(n + m + 1)$, the search space $\mathbf{S}$ of feasible values for θ is $\Re^{(n+m+1)}$.

Fig. 5.1 Adaptive IIR model for system identification

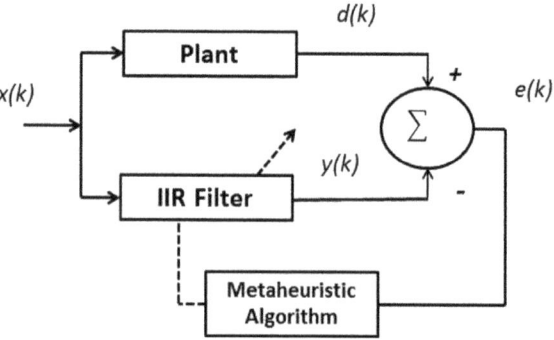

According to the diagram shown in Fig. 5.1, the output of the plant is $d(t)$ while the output of the IIR filter is represented as $y(t)$. The difference between the real system and its model produces the error $e(t) = d(t) - y(t)$. Therefore, the IIR model system identification can be considered as a minimization task as the follow expression describes:

$$f(\theta) = \frac{1}{W} \sum_{t=1}^{W} (d(t) - y(t))^2 \qquad (5.15)$$

where W is the number of samples employed for the simulation process.

The aim is to minimize the cost function $f(\theta)$ by adjusting θ. The optimal model θ^* or solution is achieved when the error function $f(\theta)$ attains its minimum value, as follows:

$$\theta^* = \arg\min_{\theta \in S}(f(\theta)) \qquad (5.16)$$

5.4 Experimental Results

For the comparative analysis, a wide set of experiments has been employed to examine the performance of each optimization technique. The experimentation examines the use of IIR systems with various orders. Such an experimental set has been meticulously chosen to guarantee compatibility between related works reported in the literature [18–21]. For the comparison, five ET's are considered: PSO, ABC, EM, CS, and FPA.

The parameter setting for the optimization techniques that are employed in the comparative study is described as follows:

1. PSO: $c_1 = 2$, $c_2 = 2$, the weight factor decreases linearly from 0.9 to 0.2 [18].
2. ABC: The algorithm was implemented using the guidelines of its own Ref. [19], $limit = 100$.
3. EM: $\delta = 0.001$, $LISTER = 4$, $MaxIter = 300$. The values, according to [20].
4. CS: The parameters are set to $p_a = 0.25$ [21].
5. FPA: The probabilistic global pollination factor p is set to 0.8 [14].

All techniques employ a population size was set to 25 ($N = 25$) while the maximum generation number was configured to 3000 ($gen = 3000$).

5.4.1 Results of IIR Model Identification

The experimental results are detailed considering three tests that involve, (1) Second-order plant by a first-order IIR system; (2) Second-order plant by a second-order IIR system; finally, (3) High-order plant by a high-order system. Every case is explained below.

(1) Second-order plant with a first-order IIR model

For this experiment, each technique is used to identify a second-order plant by a first-order IIR system. Under such considerations, the unknown plant and the IIR system operate under the following transfer functions:

$$H_P(z^{-1}) = \frac{0.05 - 0.4z^{-1}}{1 - 1.1314z^{-1} + 0.25z^{-2}}, \; H_M(z^{-1}) = \frac{b}{1 - az^{-1}} \qquad (5.17)$$

For the simulations, it was considered a sequence of 100 samples ($W = 100$) for the input $u(t)$. Since a low order system is used to identify a superior order plant, the generated multi-modal error surface is shown in Fig. 5.2 [15–17].

Fig. 5.2 Multimodal error surface $f(\theta)$ for the first experiment: **a** tridimensional figure and **b** contour

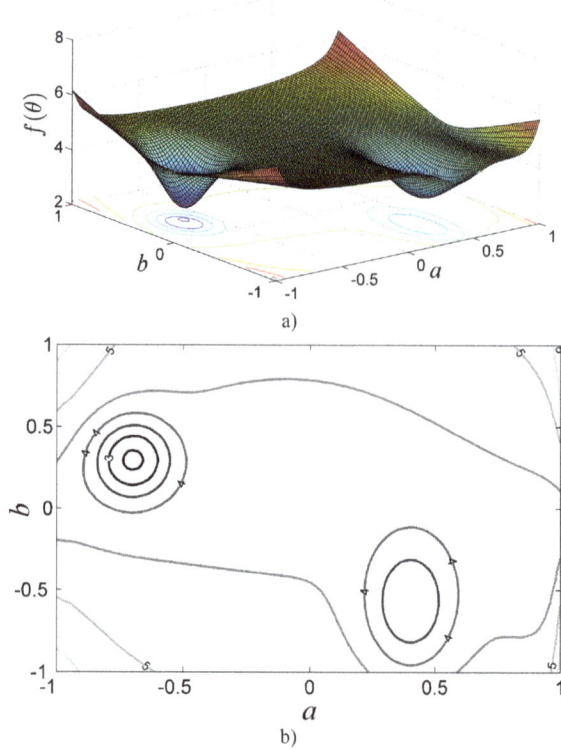

Table 5.1 Performance for the first experiment

Algorithms	ABP		AV	SD
	a	b		
PSO	0.9125	−0.3012	0.0284	0.0105
ABC	0.1420	−0.3525	0.0197	0.0015
EM	0.9034	0.3030	0.0165	0.0012
CS	0.9173	−0.2382	0.0101	3.118e−004
FPA	0.9364	−0.2001	0.0105	5.103e−004

The performance evaluation over 30 individual executions is described in Table 5.1 regarding the next indexes: best parameter design (ABP), the average $f(\theta)$ value (AV) and standard deviation (SD). The best parameter design (ABP) reports the best model parameters generated through the 30 individual executions while the average $f(\theta)$ value (AV) symbolizes the average minimum value of $f(\theta)$, regarding the same number of executions. Finally, the standard deviation (SD) describes the dispersion from the average $f(\theta)$ value concerning the 30 executions.

In Table 5.1, the CS technique presents better results than other EC's. In special, the results exhibit that CS maintains a significant accuracy (the lowest AV value) and robustness (smallest SD value). Nevertheless, the CS performance is comparable to the FPA technique. On the other hand, the PSO generates the worst performance in the experimentation. Such a fact corresponds to its problem (premature convergence) to overcome local minima over multimodal functions.

(2) Second-order plant with a system and second-order IIR system

For the second experiment, the performance for each technique is estimated for a second-order plant identification by a second-order IIR system. Hence, the unknown plant and the IIR system operate under the following transfer functions:

$$H_P(z^{-1}) = \frac{1}{1 - 1.4z^{-1} + 0.49z^{-2}}, \; H_M(z^{-1}) = \frac{b}{1 + a_1 z^{-1} + a_2 z^{-2}} \quad (5.18)$$

In the simulations, the input $u(t)$ that is employed to the model and to the IIR system simultaneously, was configured considering a sequence with 100 samples. In this experiment, the order of the system H_M is similar to the order of the system to be identified H_P, only one global minimum exists in $f(\theta)$. In the experimental results, 30 different executions are described in Table 5.2.

Table 5.2 shows that PSO, ABC, EM, CS, and FPA achieve similar results in their performance. The evidence reveals that evolutionary techniques have similar average performance when they face unimodal low-dimensional problems [28]. In general, the experiment points that the low variation in performance is directly correlated to a better exploitation mechanism involved in CS and FPA.

(3) Superior-order plant with a high-order model

Table 5.2 Performance for the second experiment

Algorithms	ABP			AV	SD
	a_1	a_2	b		
PSO	−1.4024	0.4925	0.9706	4.0035e−005	1.3970e−005
ABC	−1.2138	0.6850	0.2736	0.3584	0.1987
EM	−1.0301	0.4802	1.0091	3.9648e−005	8.7077e−005
CS	−1.400	0.4900	1.000	0.000	0.000
FPA	−1.400	0.4900	1.000	4.6246e−32	2.7360e−31

The performance for every technique is estimated considering the identification of a superior-order plant by a high-order IIR system. Hence, the unknown plant and the IIR system operate under the following transfer functions:

$$H_P(z^{-1}) = \frac{1 - 0.4z^{-2} - 0.65z^{-4} + 0.26z^{-6}}{1 - 0.77z^{-2} - 0.8498z^{-4} + 0.6486z^{-6}},$$

$$H_M(z^{-1}) = \frac{b_0 + b_1 z^{-1} + b_2 z^{-2} + b_3 z^{-3} + b_4 z^{-4}}{1 + a_1 z^{-1} + a_2 z^{-2} + a_3 z^{-3} + a_4 z^{-4}} \tag{5.19}$$

As the plant is a sixth-order model and the IIR filter is a fourth-order system, the generated error surface $f(\theta)$ tends to multimodality just as in the first test. A sequence with 100 samples was used as input. The experimental results over 30 individual executions are summarized in Tables 5.3 and 5.4. Table 5.3 shows the best parameter values (ABP) while Table 5.4 presents the average $f(\theta)$ value (AV) and the standard deviation (SD).

The AV and SD values in Table 5.4, show that the CS technique obtains best results than PSO, ABC, EM, and FPA. The results confirm that CS performs better accuracy (AV value) and robustness (SD value). These conclusions also show that CS, FPA, and EM can identify the sixth-order plant under distinct accuracy levels. On the other hand, PSO and ABC achieve sub-optimal solutions whose parameters weakly model the unknown model.

5.4.2 Statistical Study

To statistically verify the experimental results, a non-parametric statistical tool that is known as Wilcoxon's test for independent samples [24, 28], was conducted over the "the average value" (AV) results of Tables 5.1, 5.2, and 5.4 considering a 5% significance level. The experiment was carried out considering 30 individual executions for each technique. Table 5.5 summarizes the p-values generated by Wilcoxon's test for the comparative study of the average value of four algorithms. The groups are developed by CS versus PSO, CS versus ABC, CS versus EM, and CS versus

Table 5.3 Parameter values (ABP) for the third experiment

Algorithms	ABP									
	a_1	a_2	a_3	a_4	b_0	b_1	b_2	b_3	b_4
PSO	0.3683	−0.7043	0.2807	0.3818	0.9939	−0.6601	−0.8520	0.2275	−1.4990
ABC	−1.1634	−0.6354	−1.5182	0.6923	0.5214	−1.2703	0.3520	1.1816	−1.9411
EM	−0.4950	−0.7049	0.5656	−0.2691	1.0335	−0.6670	−0.4682	0.6961	−0.0673
CS	0.9599	0.0248	0.0368	−0.0002	−0.2377	0.0031	−0.3579	0.0011	−0.5330
FPA	0.0328	−0.1059	−0.0243	−0.7619	1.0171	0.0038	0.2374	0.0259	−0.3365

Table 5.4 Average $f(\theta)$ value (AV) and standard deviation (SD)

Algorithms	AV	SD
PSO	5.8843	3.4812
ABC	7.3067	4.3194
EM	0.0140	0.0064
CS	6.7515e−004	4.1451e−004
FPA	0.0018	0.0020

Table 5.5 p-values generated by Wilcoxon's test comparing CS versus PSO, ABC, EM, and FPA considering the "The average $f(\theta)$ values (AV)" from Tables 5.1, 5.2 and 5.4

CS versus	PSO	ABC	EM	FPA
First experiment	6.5455e−13	8.4673e−13	3.8593e−08	0.7870
Second experiment	1.5346e−14	1.5346e−14	1.5346e−14	0.3313
Third experiment	6.5455e−13	1.5346e−14	4.3234e−13	0.1011

FPA. The null hypothesis assumes that there is no notable difference between averaged values of the two techniques. The alternative hypothesis holds that there is a significant difference between the AV of both approaches. In the case of PSO, ABC, and EM, the p-values listed in Table 5.5 are lower than 0.05 which is strong enough evidence against the null hypothesis. Hence, this evidence shows that CS is statistically significant and that it has not determined good solutions by coincidence. On the other hand, the p-values of CS versus FPA are higher than 0.05, which means that there is no statistical difference between both techniques. Consequently, it can be assumed that the CS technique is better than PSO, ABC, and EM in the tasks of IIR models for system identification. Nevertheless, CS presents statistically the same performance that FPA and therefore there is no statistical evidence that CS overcomes the FPA technique [29].

5.5   Conclusions

This chapter presents a comparative study among five optimization techniques for the IIR-based system identification. Under this approach, the system identification task can be considered an optimization problem. In the comparative study, particular attention is given to popular techniques such as the Cuckoo Search (CS) and the Flower Pollination Algorithm (FPA), also including most used methods such as the Particle Swarm Optimization (PSO), the Artificial Bee Colony optimization (ABC) and the Electromagnetism-Like (EM) optimization techniques. The study was experimentally assessed over an experiment suite of three benchmark tests that provide multimodal functions. In the experimental results, we demonstrated that CS outperforms PSO, ABC, and EM in terms of accuracy (AV) and robustness (SD), considering a statistical framework (Wilcoxon test). Nevertheless, there is no statistical indication that CS better the FPA performance. The striking performance of CS and FPA is described by two different characteristics: (i) their exploration operators that provide a better exploration of the search space, enhancing the ability to obtain multiple optima; (ii) their exploitation operators that provide better accuracy of previously-found solutions.

References

1. Zhou X, Yang C, Gui W (2014) Nonlinear system identification and control using state transition algorithm. Appl Math Comput 226:169–179. https://doi.org/10.1016/j.amc.2013.09.055
2. Frank Pai P, Nguyen B-A, Sundaresan MJ (2013) Nonlinearity identification by time-domain-only signal processing. Int J Non Linear Mech 54:85–98. https://doi.org/10.1016/j.ijnonlinmec.2013.04.002
3. Albaghdadi M, Briley B, Evens M (2006) Event storm detection and identification in communication systems. Reliab Eng Syst Saf 91:602–613. https://doi.org/10.1016/j.ress.2005.05.001
4. Chung HC, Liang J, Kushiyama S, Shinozuka M (2004) Digital image processing for non-linear system identification. Int J Non Linear Mech 39:691–707. https://doi.org/10.1016/S0020-7462(03)00021-0
5. Na J, Ren X, Xia Y (2014) Adaptive parameter identification of linear SISO systems with unknown time-delay. Syst Control Lett 66:43–50. https://doi.org/10.1016/J.SYSCONLE.2014.01.005
6. Kukrer O (2011) Analysis of the dynamics of a memoryless nonlinear gradient IIR adaptive notch filter. Signal Process 91:2379–2394. https://doi.org/10.1016/j.sigpro.2011.05.001
7. Mostajabi T, Poshtan J, Mostajabi Z (2015) IIR model identification via evolutionary algorithms. Artif Intell Rev 44:87–101. https://doi.org/10.1007/s10462-013-9403-1
8. Dai C, Chen W, Zhu Y (2010) Seeker optimization algorithm for digital IIR filter design. IEEE Trans Ind Electron 57:1710–1718. https://doi.org/10.1109/TIE.2009.2031194
9. Fang W, Sun J, Xu W (2009) A new mutated quantum-behaved particle swarm optimizer for digital IIR filter design. EURASIP J Adv Signal Process 2009. https://doi.org/10.1155/2009/367465
10. Kennedy J, Eberhart R (1995) Particle swarm optimization. Neural Netw, 1995 Proceedings, IEEE Int Conf 4:1942–1948. https://doi.org/10.1109/ICNN.1995.488968
11. Karaboga D (2005) An idea based on honey bee swarm for numerical optimization. Tech Rep TR06, Erciyes Univ 10. https://doi.org/citeulike-article-id:6592152
12. Birbil ŞI, Fang SC (2003) An electromagnetism-like mechanism for global optimization. J Glob Optim 25:263–282. https://doi.org/10.1023/A:1022452626305
13. Yang XS, Deb S (2009) Cuckoo search via Lévy flights. 2009 World Congr Nat Biol Inspired Comput NABIC 2009—Proc 210–214. https://doi.org/10.1109/NABIC.2009.5393690
14. Yang XS (2012) Flower pollination algorithm for global optimization. Lect Notes Comput Sci (including Subser Lect Notes Artif Intell Lect Notes Bioinformatics) 7445 LNCS:240–249. https://doi.org/10.1007/978-3-642-32894-7_27
15. Ahn CW (2006) Advances in evolutionary algorithms: theory, design and practice
16. Chiong R, Weise T, Michalewicz Z (2013) Variants of evolutionary algorithms for real-world applications
17. Oltean M (2007) Evolving evolutionary algorithms with patterns. Soft Comput 11:503–518. https://doi.org/10.1007/s00500-006-0079-1

18. Serbet F, Kaya T, Ozdemir MT (2017) Design of digital IIR filter using particle swarm optimization. In: 2017 40th international convention on information and communication technology, electronics and microelectronics, MIPRO 2017—proceedings. Institute of Electrical and Electronics Engineers Inc., pp 202–204

19. Karaboga N (2009) A new design method based on artificial bee colony algorithm for digital IIR filters. J Franklin Inst 346:328–348. https://doi.org/10.1016/j.jfranklin.2008.11.003

20. Cuevas-Jiménez E, Oliva-Navarro DA (2013) Modelado de filtros IIR usando un algoritmo inspirado en el electromagnetismo. Ing Investig y Tecnol 14:125–138. https://doi.org/10.1016/s1405-7743(13)72231-5

21. Patwardhan AP, Patidar R, George NV (2014) On a cuckoo search optimization approach towards feedback system identification. Digit Signal Process A Rev J 32:156–163. https://doi.org/10.1016/j.dsp.2014.05.008

22. Wolpert DH, Macready WG (1997) No free lunch theorems for optimization. IEEE Trans Evol Comput 1:67–82. https://doi.org/10.1109/4235.585893

23. Elbeltagi E, Hegazy T, Grierson D (2005) Comparison among five evolutionary-based optimization algorithms. Adv Eng Inf 19:43–53. https://doi.org/10.1016/j.aei.2005.01.004

24. Shilane D, Martikainen J, Dudoit S, Ovaska SJ (2008) A general framework for statistical performance comparison of evolutionary computation algorithms. Inf Sci (Ny) 178:2870–2879. https://doi.org/10.1016/j.ins.2008.03.007

25. Osuna-Enciso V, Cuevas E, Sossa H (2013) A comparison of nature inspired algorithms for multi-threshold image segmentation. Expert Syst Appl 40:1213–1219. https://doi.org/10.1016/J.ESWA.2012.08.017

26. Lin Y-L, Chang W-D, Hsieh J-G (2008) A particle swarm optimization approach to nonlinear rational filter modeling. Expert Syst Appl 34:1194–1199. https://doi.org/10.1016/j.eswa.2006.12.004

27. Barthelemy P, Bertolotti J, Wiersma DS (2008) A Lévy flight for light. Nature 453:495–498. https://doi.org/10.1038/nature06948

28. García S, Molina D, Lozano M, Herrera F (2009) A study on the use of non-parametric tests for analyzing the evolutionary algorithms' behaviour: a case study on the CEC'2005 special session on real parameter optimization. J Heuristics 15:617–644. https://doi.org/10.1007/s10732-008-9080-4

29. Cuevas E, González A, Fausto F, Zaldívar D, Pérez-Cisneros M (2015) Multithreshold segmentation by using an algorithm based on the behavior of Locust Swarms. Math Probl Eng 2015:805357

Chapter 6
Fractional-Order Estimation Using via Locust Search Algorithm

Recently, parameter estimation for fractional-order chaotic systems has attracted the attention of multiple scientific and engineering fields. In the estimation process, the parameters of a given system are formulated into an optimization problem. One of the most interesting estimation problems relies on fractional-order systems. Where functional parameters and fractional orders parameters of the chaotic system are considered as decision variables in an optimization perspective. For this scenario, the complexity of fractional-order chaotic systems tends to construct error functions within a multimodal fashion. Many algorithms based on the principles of evolutionary computation have been developed to be applied to determine the parameters of a given fractional-order chaotic system. Nonetheless, most of them hold a crucial boundary, they repeatedly acquire local optimal values as outcomes, as a consequence of an improper balance among the evolutionary exploration stage and exploitation stage. This chapter introduces an evolutionary approach to estimating the parameter values of fractional-order chaotic systems. To estimate the parameters, the proposed approach employs a newly developed evolutionary technique known as Locust Search (LS) which is based on the conduct of swarms of locusts.

6.1 Introduction

Fractional-order systems are represented based on fractional differential equations incorporating derivatives of non-integer order. Many scientific and engineering fields contemplate applications of fractional-order derivatives including power lines [1], electrical power systems [2] and control systems [3]. For that, fractional order systems have increased the attention of several research areas [4–8].

The process of proper system identification of fractional-order systems relies on complex computational procedures to be evaluated. Due to the mathematical abstraction of fractional calculus, fractional derivatives are slightly different than integer derivatives [9]. Traditional integer-order derivatives can define the parameters of a

E. Cuevas et al., *Analysis and Comparison of Metaheuristics*, Studies in Computational Intelligence 1063, https://doi.org/10.1007/978-3-031-20105-9_6

given model in a direct fashion. However, for fractional-order models, the identification procedure requires more complex procedure [10]. For that, the majority of the traditional identification approaches cannot be directly employed considering a fractional-order model [11].

In general, the estimation of parameters problem considering fractional-order derivatives has been analyzed by traditional methods such as linear optimization approaches [12], input–output frequency scopes [13], and operational matrix [14]. These approaches represent the most common tools since they have been thoroughly examined and studied on the state-of-the-art [15].

Alternatively, the parameter identification problem for fractional-order models has been solved via an evolutionary computation perspective. In general, evolutionary approaches have been proved, over many complex scenarios, to produce better outcomes than those based on classical approaches in terms of robustness and accurateness [16]. In evolutionary methodologies, an individual is characterized in terms of decision variables. Then, a group of operators evaluates such decision variables through the use of biological metaphors to obtain better individuals through the generations. The quality of each individual is computed considering a cost function whose final outcome symbolizes the association among the real and approximated model. Several examples of evolutionary algorithms have been successfully applied for the identification of fractional-order models. Some of these approaches consider the Genetic Algorithms (GA) [17], Artificial Bee Colony (ABC) [18], Differential Evolution (DE) [19] and Particle Swarm Optimization (PSO) [20]. Even when these kinds of methodologies are presented as an alternative to classical methods, they have critical issues. Evolutionary approaches usually produce local optimal values as solutions due to the balance among evolutionary stages (exploration and exploitation). This restriction is related to the evolutionary operators considered to adjust the decision variables. In these algorithms, during their execution, each decision variable is modified for the following generations. Hence, as the algorithm evolves, this balance among evolutionary stages causes that the population of individuals concentrates near the best solutions, yielding premature convergence into local optima instead of global optima [21, 22].

This chapter describes an approach based on an evolutionary algorithm to correctly identify the parameters of fractional-order chaotic systems. To identify the parameters, the proposed approach employs a novel evolutionary technique named Locust Search (LS) [23–25] which is based on the characterization of locusts. In the presented methodology, the solutions (individuals of a population) mimic a set of locusts that interact among them following a set of biological laws of collective cooperation. The methodology follows two distinct conducts of locusts: social and solitary. Relying on the conduct, each solution (locust) is guided by a collection of evolutionary operators which emulate diverse collaborative behaviors that are commonly encountered in the locust swarm. Dissimilar to the majority of evolutionary techniques, the model of the presented approach overcomes the concentration of individuals. This mechanism avoids critical situations such as premature convergence. Experimental results have been evaluated considering the fractional-Order Van der Pol oscillator to demonstrate the efficacy of the proposed approach [26].

The chapter is organized as follows. Section 6.2 presents some of the basic concepts of fractional calculus. Section 6.3 describes the Locust Search approach. Section 6.4, it is presented the mathematical description of the Van der Pol Oscillator. Then, in Sect. 6.5 the parameter estimation problem of fractional-order systems is reformulated as an optimization perspective. Section 6.6 presents the experimental simulations. Finally, in Sect. 6.7 some conclusions are detailed.

6.2 Fractional Calculus

Fractional-order calculus is considered as a generalization of differentiation and integration to non-integer. The differential-integral operator, symbolized by $_aD_t^q$ carries both the fractional integral and the fractional derivative in the same mathematical model as:

$$_aD_t^q = \begin{cases} \frac{d^q}{dt^q}, & q > 0, \\ 1, & q = 0, \\ \int_a^t (d\tau)^q, & q < 0. \end{cases} \tag{6.1}$$

where a and t corresponds to the operation bounds. The general definition commonly used in fractional derivatives are the Grünwald-Letnikov, Riemann–Liouville [7] and Caputo [27]. According to the approximation method of Grünwald-Letnikov, the fractional-order derivative q is modeled as:

$$D_t^q f(t) = \lim_{h \to 0} \frac{1}{h^q} \sum_{j=0}^{\infty} (-1)^j \binom{q}{j} f(t - jh) \tag{6.2}$$

In the numerical computation of fractional-order, the numerical approximation of the qth derivative considering the values $kh, (k = 1, 2, ...)$ keeps the subsequent form [28]:

$$_{(k-L_m/h)}D_{t_k}^q f(t) \approx h^{-q} \sum_{j=0}^{k} (-1)^j \binom{q}{j} f(t_k - j) \tag{6.3}$$

where L_m represents the memory length term $t_k = kh, h$, corresponds to the time step, and $(-1)^j \binom{q}{j}$ are the binomial coefficients. For their computation it is possible to apply the following description:

$$c_0^{(q)} = 1, \quad c_j^{(q)} = \left(1 - \frac{1+q}{j}\right)c_{j-1}^{(q)} \tag{6.4}$$

Finally, the general numerical solution of the fractional differential equation is determined as:

$$y(t_k) = f(y(t_k), t_k)h^q - \sum_{j=1}^{k} c_j^{(q)} y(t_{k-j}) \tag{6.5}$$

6.3 Locust Search (LS) Algorithm

In the algorithm procedure of LS [23], the population of N solutions (locusts) $\mathbf{L}^k$ ($\{\mathbf{l}_1^k, \mathbf{l}_2^k, \ldots, \mathbf{l}_N^k\}$) is operated from initial generation towards a maximum number of generations. Each locust $\mathbf{l}_i^k$ ($i \in [1, \ldots, N]$) is coded based on an n-dimensional decision variable vector $\{l_{i,1}^k, l_{i,2}^k, \ldots, l_{i,n}^k\}$. The group of these decision vector span into the search space $\mathbf{S} = \{\mathbf{l}_i^k \in \mathbb{R}^n | lb_d \leq l_{i,d}^k \leq ub_d\}$, where lb_d and ub_d corresponds to the boundaries of the search space. The fitness value of each locust $\mathbf{l}_i^k$ refers to the quality of the solution and its value corresponds to the evaluation of the cost function $f(\mathbf{l}_i^k)$. In the algorithmic structure of LS, each generation two evolutionary operators are performed: (A) solitary phase and (B) social phase.

6.3.1 Solitary Phase (A)

Regarding to the solitary phase, a new position of a locust $\mathbf{p}_i$ ($i \in [1, \ldots, N]$) is modified considering a change of position $\Delta \mathbf{l}_i$ ($\mathbf{p}_i = \mathbf{l}_i^k + \Delta \mathbf{l}_i$). $\Delta \mathbf{l}_i$ is the combination of the solution cooperation by $\mathbf{l}_i^k$ as their biological conduct. This cooperation are computed pairwise between $\mathbf{l}_i^k$ and the remaining locusts (solutions). Hence, the attraction among $\mathbf{l}_j^k$ and $\mathbf{l}_i^k$ is emulated considering:

$$\mathbf{s}_{ij}^m = \rho(\mathbf{l}_i^k, \mathbf{l}_j^k) \cdot s(r_{ij}) \cdot \mathbf{d}_{ij} + rand(1, -1) \tag{6.6}$$

where $\mathbf{d}_{ij} = (\mathbf{l}_j^k - \mathbf{l}_i^k)/r_{ij}$ corresponds to the unit-vector in the direction of $\mathbf{l}_i^k$ to $\mathbf{l}_j^k$. Also, the expression $rand(1, -1)$ is considered as a uniform random value from the range $[-1\ 1]$. The element $s(r_{ij})$ exposes the social association among $\mathbf{l}_j^k$ and $\mathbf{l}_{.i}^k$:

$$s(r_{ij}) = F \cdot e^{-r_{ij}/L} - e^{-r_{ij}} \tag{6.7}$$

where, r_{ij} is considered as the distance among the solutions $\mathbf{l}_j^k$ and $\mathbf{l}_{.i}^k$, F expresses the attraction effect and L is length step of the attraction. It is observable that if $F < 1$ and $L > 1$ the repulsion force is stronger contrary to this, the attraction is weaker. $\rho(\mathbf{l}_i^k, \mathbf{l}_j^k)$ corresponds to the evaluation of dominance among $\mathbf{l}_j^k$ and $\mathbf{l}_i^k$. To evaluate

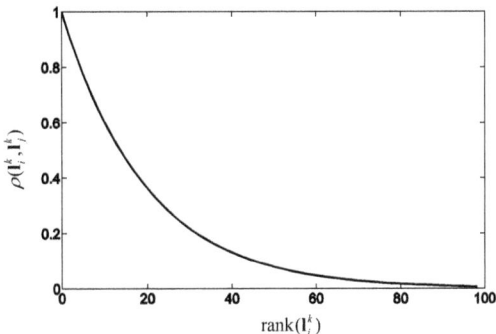

Fig. 6.1 Characterization of $\rho(\mathbf{l}_i^k, \mathbf{l}_j^k)$ with 100 elements

the function $\rho(\mathbf{l}_i^k, \mathbf{l}_j^k)$, the individuals of the population $\mathbf{L}^k(\{\mathbf{l}_1^k, \mathbf{l}_2^k, \ldots, \mathbf{l}_N^k\})$ are sorted according to the quality of the solutions. Then, a rank value is determined for each locust, the best solution so far acquires the rank zero while the worst locust obtains the rank $N - 1$. The dominance function $\rho(\mathbf{l}_i^k, \mathbf{l}_j^k)$ is:

$$\rho(\mathbf{l}_i^k, \mathbf{l}_j^k) = \begin{cases} e^{-(5 \cdot \text{rank}(\mathbf{l}_i^k)/N)} & \text{if } \text{rank}(\mathbf{l}_i^k) < \text{rank}(\mathbf{l}_j^k) \\ e^{-(5 \cdot \text{rank}(\mathbf{l}_j^k)/N)} & \text{if } \text{rank}(\mathbf{l}_i^k) > \text{rank}(\mathbf{l}_j^k) \end{cases} \tag{6.8}$$

where $\text{rank}(\alpha)$ refers to the rank for the α-element, $\rho(\mathbf{l}_i^k, \mathbf{l}_j^k)$ computes the dominance within the interval $[0, 1]$. Figure 6.1 presents the characterization of $\rho(\mathbf{l}_i^k, \mathbf{l}_j^k)$ regarding 100 points. In the figure, $\mathbf{l}_i^k$ corresponds to one of the 99 elements.

Then, the obtain force $\mathbf{S}_i^m$ over each locust (solution) $\mathbf{l}_i^k$ is computed following the superposition principal over pairwise cooperation as:

$$\mathbf{S}_i^m = \sum_{\substack{j=1 \\ j \neq i}}^{N} \mathbf{s}_{ij}^m \tag{6.9}$$

Finally, $\Delta \mathbf{l}_i$ represents the social force exerted by $\mathbf{l}_i^k$. As a result, $\Delta \mathbf{l}_i$ is computed as:

$$\Delta \mathbf{l}_i = \mathbf{S}_i^m \tag{6.10}$$

After obtaining the new positions of the solutions in the population $\mathbf{P}(\{\mathbf{p}_1, \mathbf{p}_2, \ldots, \mathbf{p}_N\})$, the final locations $\mathbf{F}(\{\mathbf{f}_1, \mathbf{f}_2, \ldots, \mathbf{f}_N\})$ are required to be calculated. This procedure is achieved by the following model:

$$\mathbf{f}_i = \begin{cases} \mathbf{p}_i & \text{if } f(\mathbf{p}_i) < f(\mathbf{l}_i^k) \\ \mathbf{l}_i^k & \text{otherwise} \end{cases} \tag{6.11}$$

6.3.2 Social Phase (B)

In this stage of the LS method, considers a discriminant operation of the locust swarm called social phase. This operation considers to a collection $\mathbf{E}$ of the final locations $\mathbf{F}$ (where $\mathbf{E} \subseteq \mathbf{F}$). In this strategy, by initial state the elements of $\mathbf{F}$ are sorted based on their quality values (fitness function value) in a temporal population set $\mathbf{B} = \{\mathbf{b}_1, \mathbf{b}_2, \ldots, \mathbf{b}_N\}$. The elements inside this temporal population are arranged so that the best individual is at the first position $\mathbf{b}_1$ $\{b_{1,1}, b_{1,2}, \ldots, b_{1,n}\}$ while the worst solution is in the last location $\mathbf{b}_N$. For that, $\mathbf{E}$ contains by the first g location of $\mathbf{B}$. Later, a subspace C_j is determined over $\mathbf{f}_j \in \mathbf{E}$. The size of C_j is based on the distance e_d which is calculated as:

$$e_d = \frac{\sum_{q=1}^{n} \left(ub_q - lb_q\right)}{n} \cdot \beta \qquad (6.12)$$

where ub_q and lb_q are the search space limits, n corresponds to the number of dimensions, $\beta \in [0, 1]$ is a tuning variable. Hence, the limit boundaries of C_j are calculated as:

$$\begin{aligned} uss_j^q &= b_{j,q} + e_d \\ lss_j^q &= b_{j,q} - e_d \end{aligned} \qquad (6.13)$$

where uss_j^q and lss_j^q are the boundary limits of the subspace C_j. Once constructed the subspace C_j around the $\mathbf{f}_j \in \mathbf{E}$, a collection of h individuals ($\mathbf{M}_j^h = \{\mathbf{m}_j^1, \mathbf{m}_j^2, \ldots, \mathbf{m}_j^h\}$) are produced in a random fashion considering Eq. 6.13. Taking into account the h individuals, the new solution $\mathbf{l}_j^{k+1}$ of the future population $\mathbf{L}^{k+1}$ must be constructed. For that, the best element $\mathbf{m}_j^{best}$ is analyzed. If $\mathbf{m}_j^{best}$ is better than $\mathbf{f}_j$, $\mathbf{l}_j^{k+1}$ is modified with $\mathbf{m}_j^{best}$, otherwise $\mathbf{f}_j$ is assigned to $\mathbf{l}_j^{k+1}$. The social phase is considered as an exploitation strategy in the LS method.

6.4 Fractional-Order Van der Pol Oscillator

The Van der Pol Oscillator system has been extensively analyzed in several research fields. It is considered as a complex chaotic system containing of non-linear nature. It provides interesting systems with dynamic behaviors for several applications in science and engineering [29, 30]. The traditional integer-order Van der Pol Oscillator is mathematically determined by a second-order differential equation as:

$$\begin{bmatrix} \dot{y}_1 \\ \dot{y}_2 \end{bmatrix} = \begin{bmatrix} 0 & 1 \\ -1 & -\varepsilon(y_1^2(t) - 1) \end{bmatrix} \begin{bmatrix} y_1 \\ y_2 \end{bmatrix}, \qquad (6.14)$$

where ε represents a control parameter that incorporates the non-linearity degree of the entire model. Also, the fractional-order Van der Pol Oscillator system of order q is mathematically described as [31]:

$$_0 D_t^{q_1} y_1(t) = y_2(t),$$
$$_0 D_t^{q_2} y_2(t) = -y_1(t) - \varepsilon(y_1^2(t) - 1)y_2(t). \tag{6.15}$$

Taking into account the Grünwald-Letnikov approximation (Eq. 6.5) for fractional-order system, the numerical solution for the fractional-order version of the well-known Van der Pol Oscillator is modeled as follows:

$$y_1(t_k) = y_2(t_{k-1})h^{q_1} - \sum_{j=1}^{k} c_j^{(q_1)} y_1(t_{k-j}),$$

$$y_2(t_k) = (-y_1(t_k) - \varepsilon(y_1^2(t_k) - 1)y_2(t_{k-1}))h^{q_2} - \sum_{j=1}^{k} c_j^{(q_2)} y_2(t_{k-j}). \tag{6.16}$$

6.5 Problem Formulation

In the presented research in this chapter, the identification problem of estimating parameters over fractional-order systems has been reformulated considering the multimodal and multidimensional complexity of the optimization problem. For the optimization procedure, the decision variable vector of a fractional-order chaotic system (symbolized as FOC_E) is estimated through the evolutionary scheme of the LS algorithm The main goal is that FOC_E presents the best matching with FOC_O. For that, the original FOC_O fractional-order chaotic system can be mathematically computed as the following model:

$$_a D_t^q Y = F(Y, Y_0, \theta), \tag{6.17}$$

where $Y = [y_1, y_2, ..., y_m]^T$ represents the state vector of the model, Y_0 corresponds to the initial vector of states, $\theta = [\theta_1, \theta_2, ..., \theta_m]^T$ symbolizes the original collection of parameter, and $q = [q_1, q_2, ..., q_m]^T$ for $0 < q_i < 1$ ($i \in [1, ..., m]$) is the vector of the fractional derivatives. Also, F is a generic non-linear relation. The approximated fractional-order chaotic system FOC_E can be mathematically modeled according to the following structure:

$$_a D_t^{\hat{q}} \hat{Y} = F(\hat{Y}, Y_0, \hat{\theta}), \tag{6.18}$$

where $\hat{\mathbf{Y}}$, $\hat{\boldsymbol{\theta}}$ and $\hat{\mathbf{q}}$ represents the approximated system, the approximated parameter vector and the approximated fractional orders.

Since the main idea of this approach is that FOC_E presents the best approximation to FOC_O, the parameter estimation problem can be reformulated into as an optimization problem as follows:

$$\overline{\theta}, \overline{q} = \arg \min_{(\hat{\mathbf{Y}}, \hat{q}) \in \Omega} (J(\theta, q)), \tag{6.19}$$

where $\overline{\theta}, \overline{q}$ indicates the best candidate vector of parameters acquired by the optimization model, Ω represents the inherent search space while the J represents the optimization cost function which models the approximation error among FOC_O and FOC_E. The objective function is presented as:

$$J(\theta, q) = \frac{1}{M} \sum_{k=1}^{M} (\mathbf{Y}(k) - \hat{\mathbf{Y}}(k))^2, \tag{6.20}$$

where $\mathbf{Y}(k)$ and $\tilde{\mathbf{Y}}(k)$ corresponds to the original and estimated systems, respectively. k represents sampling time and M corresponds to the number of samples. Regarding to the optimization formulation in Eq. 6.19, Fig. 6.2 graphically presents parameter estimation representation of the identification process. In the figure, it is exposed the relationship among the evolutionary method and the optimization formulation.

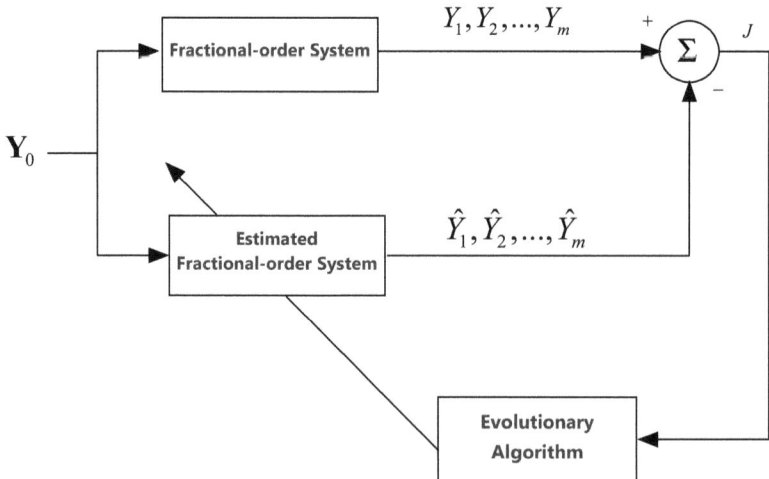

Fig. 6.2 Flowchart for fractional-order parameter approximated using evolutionary methods

6.6 Experimental Results

In this section, the performance of the proposed approach in terms of robustness and effectiveness is analyzed. The experimental scenarios have been achieved considering MATLAB (Version 7.1, MathWorks, Natick, MA, USA) on an Intel(R) Core(TM) i7-3470 CPU, 3.2 GHz with 4 GB of RAM.

The numerical results in the performance comparison among the LS and the Genetic Algorithms (GA) [17], Particle Swarm Optimization (PSO) method [20], the Differential Evolution (DE) [19] is presented in this section. For comparison purposes, the size of the population has been configured as ($N = 40$). And the number of iterations has been set to 100. This condition has been chosen to preserve compatibility with published literature [16].

The initial parameter configuration for each evolutionary method is as follows:

1. GA: The crossover probability equal 0.55, the mutation probability equal 0.10 and number of parents as 2. Additionally, the chosen selection strategy is the roulette wheel.
2. PSO: Both parameters are equal 2. ($c_1 = c_2 = 2$) and the inertia factor (ω) is linearly decreased from [0.9–0.2].
3. DE: The DE/rand/1 variation is considered. The parameter configuration values are set according to [32]. The weighting factor is $F = 0.8$ and the crossover probability is $CR = 0.9$.
4. In LS, both parameters F and L are equal to 0.6. Similarly, g is set to 20 ($N/2$), $h = 2$, $\beta = 0.6$.

In the experimental simulations, the Van der Pol Oscillator model considers $q_1 = 1.2$, $q_2 = 0.8$ and $\varepsilon = 1$ with the initial state [0.02, −0.2].

To statistically corroborates the performance results, the mean and the worst estimated parameters are evaluated considering 100 executions. The numerical results are presented in Table 6.1. From Table 6.1, it can be shown that the approximated values obtained by LS are closer to the original parameter values, which indicates that it is more accurate than the rest of tested methods. Likewise, the relative error values generated by LS are better than the rest of methods, which suggests that the proposed method obtains better accuracy in the approximation of the model. With this proof, it can be deduced that LS can estimate the fractional-order systems in a better way than its competitors. To show the proficiency of LS, Fig. 6.3 exhibits the phase diagrams of the fractional-order Van der Pol Oscillator considering the mean value.

The convergence curves of the cost function values and the system parameter values for the evolutionary methods are presented in Figs. 6.4, 6.5 and 6.6. According to the figures, it can be shown that convergence of the fitness values and parameter values of LS, present better performance than the rest of the methodologies. Also, LS converges faster.

Table 6.2 depicts the average best solution acquired by the evolutionary methods. The average best solution (ABS) describes the average value of the cost function in

Table 6.1   Performance results of the evolutionary approaches

	Parameter	GA	PSO	DE	LS
BEST	ε	0.9021	0.9152	0.9632	0.9978
	$\frac{|\varepsilon-1|}{1}$	0.0979	0.0848	0.0368	0.0022
	q_1	1.3001	1.2810	1.2210	1.2005
	$\frac{|q_1-1.2|}{1.2}$	0.0834	0.0675	0.0175	0.0004
	q_2	0.8702	0.8871	0.8229	0.8011
	$\frac{|q_2-0.8|}{0.8}$	0.0877	0.1088	0.0286	0.0013
WORST	ε	0.1731	0.1176	0.3732	0.7198
	$\frac{|\varepsilon-1|}{1}$	0.8269	0.8824	0.6268	0.2802
	q_1	2.1065	0.3643	1.8532	1.3075
	$\frac{|q_1-1.2|}{1.2}$	0.7554	0.6964	0.5443	0.0895
	q_2	0.1219	1.7643	1.2154	0.9101
	$\frac{|q_2-0.8|}{0.8}$	0.8476	1.2053	0.5192	0.1376
MEAN	ε	1.2131	1.2052	1.1701	1.0186
	$\frac{|\varepsilon-1|}{1}$	0.2131	0.2052	0.1701	0.0186
	q_1	0.9032	1.0974	1.3421	1.2654
	$\frac{|q_1-1.2|}{1.2}$	0.2473	0.0855	0.1186	0.0545
	q_2	0.9052	0.7229	0.7832	0.8089
	$\frac{|q_2-0.8|}{0.8}$	0.1315	0.0963	0.0210	0.0111

the entire set of 100 iterations. The statistical significance test Wilcoxon's rank sum has been applied [33, 34] with the 5% of the significance level. Table 6.3 reports the p-values obtained by the non-parametric test using the ABS metric among each pair of evolutionary methods. The considered null hypothesis in the study assumes that there is no significant difference among a given pair of methods. On the other hand, the alternative hypothesis assumes a significant difference among methodologies. All the reported p-values in the table are less than 5% of significance which indicates that LS obtained statistically different results.

6.7   Conclusions

The parameter estimation process is an important and interesting research field that is found in several applications, parameter identification for fractional-order systems is an active research field in several scientific and engineering groups. In the estimation process, the process of approximation is formulated as an optimization task where fractional orders are considered as the vector of decision variables in the optimization problem. Under this scheme, the complexity of fractional-order models generates

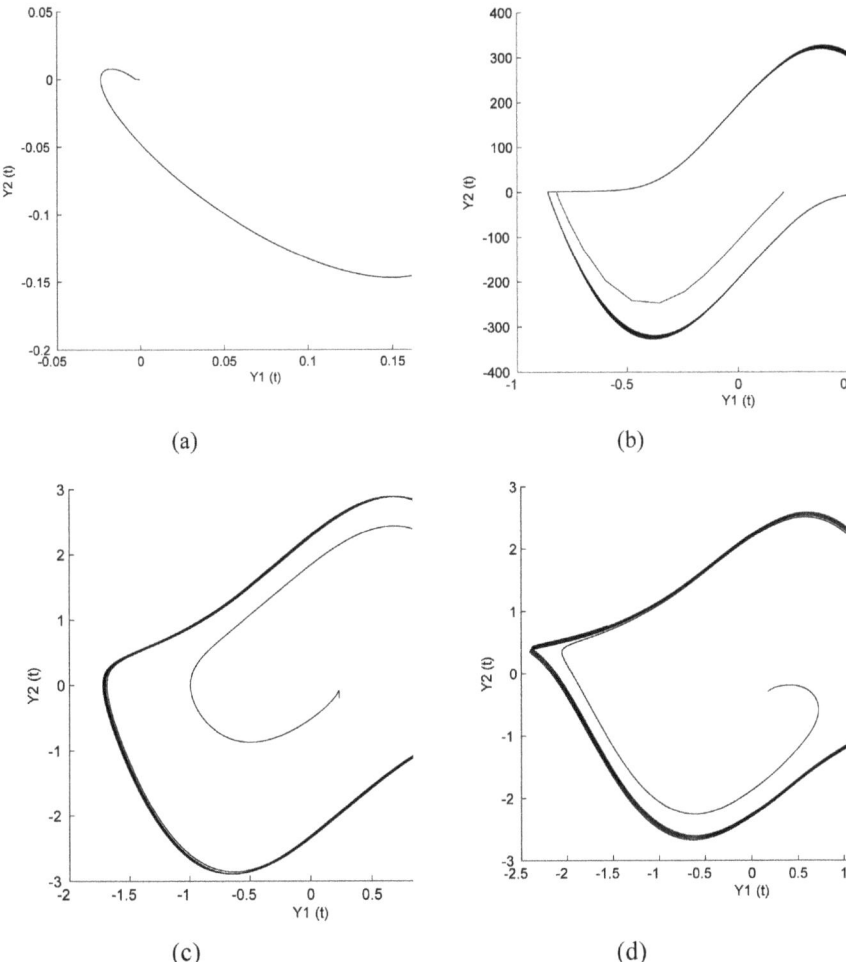

Fig. 6.3 Phase diagrams of the Van der Pol Oscillator considering the mean values of **a** GA, **b** PSO, **c** DE and **d** LS method

error surfaces containing multiple local optima. For that, traditional optimization methods are prone to fail. Recently, evolutionary methods have been proposed to estimate the parameter values of fractional-order models. Nevertheless, most of them keep an important limitation, they repeatedly acquire sub-optimal solutions. For that, non-adequate balance among evolutionary stages is an important consideration to correctly identify fractional-order models.

In this chapter, a recently developed evolutionary method Locust Search (LS) is presented to improve the estimation process of fractional-order models. The metaphor of the LS is based on the peculiar behavior of locusts. In the proposed approach, each individual (locust) cooperates with each other based on the behavior of solitary and

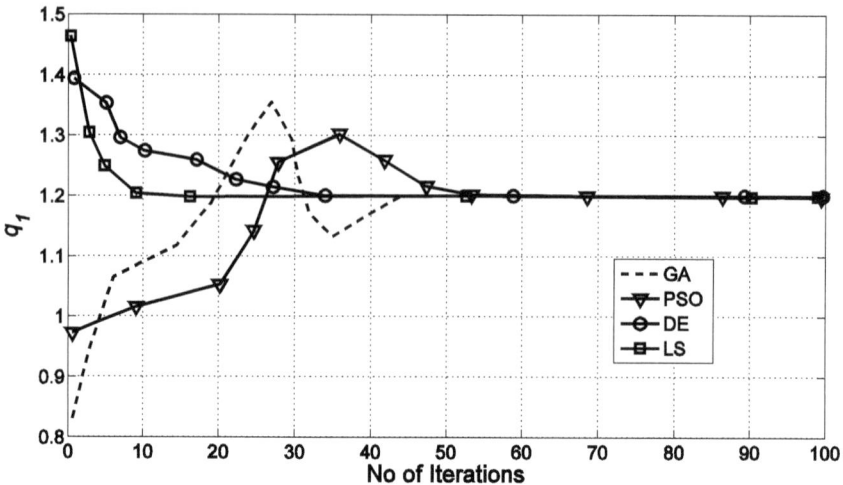

Fig. 6.4 Approximated parameter q_1 (fractional order)

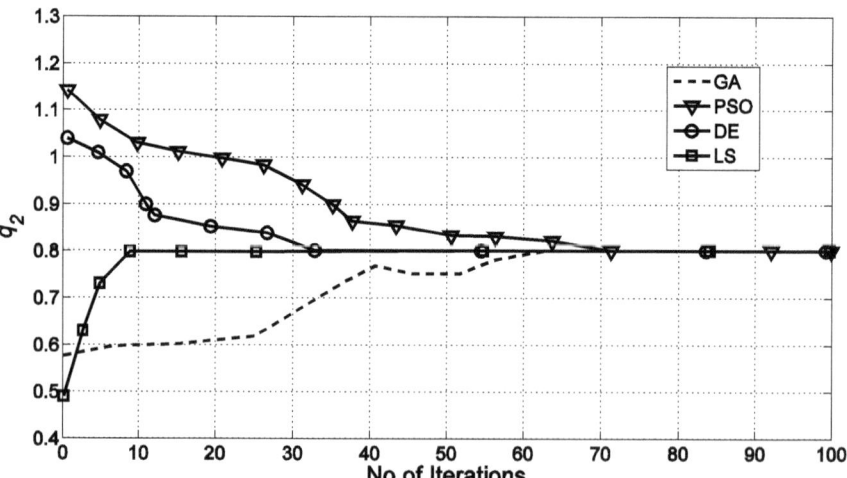

Fig. 6.5 Approximated parameter q_2 (fractional order)

social phases of the swarm of locusts. These two evolutionary operators generate the search strategy to correctly identify the parameters of fractional-order systems. Depending on the behavior exposed by each locust (solution), the cooperative inter-actions among locusts avoid the concentration of locusts in potential search zones. This feature avoids local optima stagnation in the entire search space and allows a well-balanced among evolutionary stages during the optimization process.

To test the performance of the proposed method, the chapter includes an experi-mental study to measure the robustness and accuracy of the proposed method against

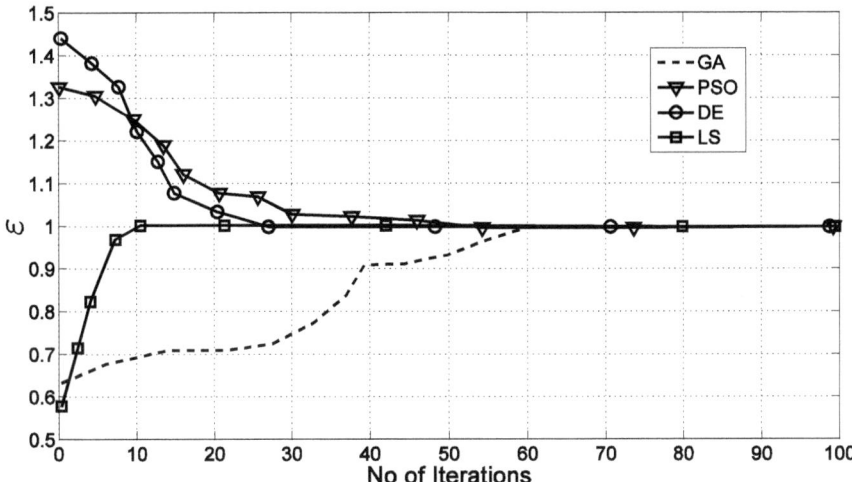

Fig. 6.6 Approximated systematic parameter ε

Table 6.2 ABS metric for each evolutionary approach

GA	PSO	DE	LS
0.2251	0.2016	0.0982	0.0126

Table 6.3 p-values obtained by the non-parametric test

	p-values
LS versus GA	0.00021
LS versus PSO	0.00098
LS versus DE	0.00123

other evolutionary methods commonly applied to solve complex optimization tasks. In the comparison study, the fractional-order chaotic system of the Van der Pol Oscillator is examined. Performance results indicate that LS is able to correctly identify the parameters values of the system in a better way than the rest of the tested evolutionary approaches.

References

1. Das S (2011) Observation of fractional calculus in physical system description. Springer, New York, pp 101–156
2. Arena P, Caponetto R, Fortuna L, Porto D (2000) Nonlinear noninteger order circuits and systems—an introduction. World Scientific, Singapore, Singapore
3. Rivero M, Rogosin SV, Tenreiro Machado JA, Trujillo JJ (2013) Stability of fractional order systems. Math Probl Eng 2013. https://doi.org/10.1155/2013/356215

4. Diethelm K (2011) An efficient parallel algorithm for the numerical solution of fractional differential equations. Fract Calc Appl Anal 14(3):475–490
5. Diethelm K, Ford NJ (2002) Analysis of fractional differential equations. J Math Anal Appl 265(2):229–248
6. Kilbas AAA, Srivastava HM, Trujillo JJ (2006) Theory and applications of fractional differential equations. Elsevier Science
7. Podlubny I (1998) Fractional differential equations. Academic Press
8. Miller KS, Ross B (1993) An introduction to the fractional calculus and fractional differential equations. Wiley, New York
9. Hu W, Yu Y, Zhang S. A hybrid artificial bee colony algorithm for parameter identification of uncertain fractional-order chaotic systems. Nonlinear Dyn. https://doi.org/10.1007/s11071-015-2251-6
10. Yu Y, Li H-X, Wang S, Yu J (2009) Dynamic analysis of a fractional-order Lorenz chaotic system. Chaos Solitons Fractals 42:1181–1189
11. Petras I (2008) Fractional-order nonlinear systems
12. Poinot T, Trigeassou J-C (2004) Identification of fractional systems using an output error technique. Nonlinear Dyn 38:133–154
13. Nazarian P, Haeri M, Tavazoei MS (2010) Identifiability of fractional order systems using input output frequency contents. ISA Trans 49:207–214
14. Saha RS (2012) On Haar wavelet operational matrix of general order and its application for the numerical solution of fractional Bagley Torvik equation. Appl Math Comput 218:5239–5248
15. Kerschen G, Worden K, Vakakis AF, Golinval JC (2006) Past, present and future of nonlinear system identification in structural dynamics. Mech Syst Signal Process 20(3):505–592
16. Quaranta G, Monti G (2010) Giuseppe Carlo Marano parameters identification of Van der Pol-Duffing oscillators via particle swarm optimization and differential evolution. Mech Syst Signal Process 24:2076–2095
17. Zhou S, Cao J, Chen Y (2013) Genetic algorithm-based identification of fractional-order systems. Entropy 15:1624–1642
18. Hu W, Yu Y, Wang S (2015) Parameters estimation of uncertain fractional-order chaotic systems via a modified artificial bee colony algorithm. Entropy 17:692–709. https://doi.org/10.3390/e17020692
19. Gao F, Lee X, Fei F, Tong H, Deng Y, Zhao H (2014) Identification time-delayed fractional order chaos with functional extrema model via differential evolution. Expert Syst Appl 41(4):1601–1608
20. Wu D, Ma Z, Li A, Zhu Q (2011) Identification for fractional order rational models based on particle swarm optimization. Int J Comput Appl Technol 41(1/2):53–59
21. Tan KC, Chiam SC, Mamun AA, Goh CK (2009) Balancing exploration and exploitation with adaptive variation for evolutionary multi-objective optimization. Eur J Oper Res 197:701–713
22. Chen G, Low CP, Yang Z (2009) Preserving and exploiting genetic diversity in evolutionary programming algorithms. IEEE Trans Evol Comput 13(3):661–673
23. Cuevas E, González A, Fausto F, Zaldívar D, Pérez-Cisneros M (2015) Multithreshold segmentation by using an algorithm based on the behavior of locust swarms. Math Probl Eng 2015, Article ID 805357, 25. https://doi.org/10.1155/2015/805357
24. Cuevas E, Zaldivar D, Perez M (2016) Automatic segmentation by using an algorithm based on the behavior of locust swarms. In: Applications of evolutionary computation in image processing and pattern recognition. Volume 100 of the series Intelligent systems reference library, pp 229–269
25. Cuevas E, González A, Zaldívar D, Pérez-Cisneros M (2015) An optimisation algorithm based on the behaviour of locust swarms. Int J Bio-Inspir Comput 7(6):402–407
26. Cuevas E, González A, Fausto F, Zaldívar D, Pérez-Cisneros M (2015) Multithreshold segmentation by using an algorithm based on the behavior of locust swarms. Math Probl Eng 2015:805357
27. Miller KS, Ross B (1993) An introduction to the fractional calculus and fractional differential equations. Wiley

28. Dorcak L (1994) Numerical models for the simulation of the fractional-order control systems
29. Quaranta G, Monti G, Marano GC (2010) Parameters identification of Van der Pol-Duffing oscillators via particle swarm optimization and differential evolution. Mech Syst Signal Process 24(7):2076–2095
30. Barbosa RS, Machado JAT, Vinagre BM, Calderon AJ (2007) Analysis of the Van der Pol oscillator containing derivatives of fractional order. J Vib Control 13(9–10):1291–1301
31. Cartwright J, Eguiluz V, Hernandez-Garcia E, Piro O (1999) Dynamics of elastic excitable media. Int J Bifur Chaos Appl Sci Eng 2197–2202
32. Cuevas E, Zaldivar D, Pérez-Cisneros M, Ramírez-Ortegón M (2011) Circle detection using discrete differential evolution optimization. Pattern Anal Appl 14(1):93–107
33. Wilcoxon F (1945) Individual comparisons by ranking methods. Biometrics 1:80–83
34. Garcia S, Molina D, Lozano M, Herrera F (2008) A study on the use of non-parametric tests for analyzing the evolutionary algorithms' behaviour: a case study on the CEC'2005 special session on real parameter optimization. J Heurist. https://doi.org/10.1007/s10732-008-9080-4

Chapter 7
Comparison of Optimization Techniques for Solar Cells Parameter Identification

The management of renewable energies has increased over the years due to the environmental effects that fossil fuels cause. Several energies alternatives have been adopted for their exploitation recently, one of the most accepted is the solar cells due to their unlimited source of power. Solar cells parameter determination has become an important task for several fields in science. The performance of solar cells operation is associated with the parameters of their design, which becomes a complex task due to the multimodality and non-linearity of their error surfaces, increasing the difficulty of their optimization. EC's techniques are employed for determining optimal solutions to complex problems. Nevertheless, when a new proposal of ECT is produced, it must be tested applying well-known functions with a specific solution without studying real-world applications which usually tend to the non-linearity and with unknown behavior. Several EC's have been employed for determining the parameters of the solar cells. Nevertheless, there is a crucial weakness that all techniques share: Generally, they achieve sub-optimal solutions due to the inappropriate equilibrium between exploration and exploitation in the search strategy. In this chapter, a comparison of optimization techniques for solar cells parameter identification is proposed, applying the one diode, two diode, and three diode models. This work performs three solar cell models considering different operation conditions, the experimental results are presented and statistically validated.

7.1 Introduction

The increasing demand and deficiency of fossil fuels [1], global warming, air pollution, and many environmental impacts have demanded that scientists explore alternative energy sources. Solar energy is the most promising renewable alternative energy souse due to its endless source of power. Solar photovoltaic energy has been widely studied around the world in several fields of engineering, increasing its use over the years [2], particularly for the free-emission generation [3], simple maintenance, and also being a reliable option in remote areas.

© The Author(s), under exclusive license to Springer Nature Switzerland AG 2023
E. Cuevas et al., *Analysis and Comparison of Metaheuristics*, Studies in Computational Intelligence 1063, https://doi.org/10.1007/978-3-031-20105-9_7

Solar cells modeling is a difficult problem due to the high non-linearity of the current versus voltage curve, moreover, its dimensionality. Solar cells are sensitive to varying operating conditions, as the temperature of the cell, the partially shaded conditions [4], just to mention a few. One of the most efficient methods for solar cells modeling is considering an electrical equivalent model, in the state-of-art are reported two different models: the single and the double diode model. In the single diode approach, just five design parameters are employed [5], while in the double diode model, seven design parameters are determined [6], which complicates its modeling. The third model considered for the study is the three-diode model which was recently introduced [7], this model employs ten different design parameters.

Some methods have been reported in literature for the modeling solar cells, such as the solar cell parameters extraction based on Newton's method [8], a method for the parameters estimation analytically using the Lambert function [9], and some other methods that use analytical approaches to determinate the parameters for the solar cell modeling [10, 11]. These methods can model the solar cells with a good precision, but their implementation become tedious and inefficient, for this, alternative techniques have been proposed to solve this problem competitively. Traditional optimization methods for the solar cells modeling such as gradient-based [12], have been used for parameters extraction in some applications, although, these techniques are susceptible to find sub-optimal solutions due their limitations in its search strategy and the error surface generated by the complex applications.

Different techniques have been presented in the literature for solar cells modeling as the solar cell parameters estimation using Newton's technique [8], a process for the estimation of the parameters analytically using the Lambert function [9], and some different approaches that apply analytical methods to estimate the solar cell parameters [10, 11]. These systems can model the solar cells with good accuracy, but their implementation is a tedious and inefficient task, for this, different methods were introduced for solving this problem competitively. Classical optimization techniques for solar cells modeling as gradient-based [12] have been applied for parameters determination in specific applications, although, these procedures are susceptible to determine sub-optimal solutions due to their weaknesses in its search strategy and considering the error surface generated by the difficult problems.

EC's approaches are optimizations techniques employed to solve complex engineering problems due to the operators used in their search strategy, which usually can avoid the local minima, determining optimal solutions. Many EC's techniques and their modifications have been introduced for the solar cell parameters determination such as the Genetic Algorithms (GA) [13, 14], in which the parameter estimation is focused on the single diode equivalent circuit, and the experimental results are summarized individually, the Particle Swarm Optimization (PSO) [15, 16], Artificial Bee Colony (ABC) [17, 18], Differential Evolution (DE) [19, 20], Harmony Search (HS) [21, 22], Cuckoo Search (CS) [23, 24], just for mention a few, are employed to estimate the solar cell parameters such as single as double diode model. The parameter estimation considering the single and double diode model [25–27]. On the other hand, the three-diode model has been poorly informed [7]. The three diode models have been designed applying different procedures described in the literature, but a comparative study between them has not been validated properly.

In this chapter, a comparative study of popular optimization approaches for parameter determination of single, double, and three diode model solar cells is presented. For this study, the Artificial Bee Colony (ABC) [28], Differential Evolution (DE) [29], Harmony Search (HS) [30], Gravitational Search Algorithm (GSA) [31], Cuckoo Search (CS) [32], Differential Search Algorithm (DSA) [33], the Crow Search Algorithm (CSA) [34], and Covariant Matrix Adaptation with Evolution Strategy (CMA-ES) [35, 36], are employed. All these techniques have shown competitive performance in benchmark test functions. Since the solar cells parameter estimation error surface tends to multimodality, moreover, the non-linearity, the dimensionality, and the difficulty of the application, a non-parametric test is included to verify the performance of the experimental results.

7.2 Evolutionary Computation (EC) Techniques

Evolutionary computation (EC) techniques are helpful mechanisms that are employed for solving difficult problems with high accuracy. These techniques are developed to solve particular tasks with particular characteristics, under such circumstances, a single technique cannot solve all problems efficiently, to overcome this, different optimization techniques are developed over well-known benchmark functions with an exact solution without taking into account real problems which normally tend to generate multimodal surfaces and high dimensionality. The main obstacle when a new optimization technique is developed, is the accurate valance between the exploration and exploitation for the search strategy. For this, a study for the best method for a specific task must be generated, and the experimental results obtained must be statistically validated to verify the accuracy and consistency of the approaches.

7.2.1 Artificial Bee Colony (ABC)

The ABC was introduced by Karaboga [28], it is based on the intelligent foraging performance of the honeybee swarm. From the computational point of view, the ABC algorithm produces a population $\mathbf{X}^k(\{\mathbf{x}_1^k, \mathbf{x}_2^k, \ldots, \mathbf{x}_N^k\})$ of N potential food positions, which are evolved from the initial point ($k = 0$) to stop criterion. Each possible food location $\mathbf{X}_i^k (i \in [1, \ldots, N])$ symbolizes a d-dimensional vector $\{\mathbf{x}_{i,1}^k, \mathbf{x}_{i,2}^k, \ldots, \mathbf{x}_{i,d}^k\}$ where the decision variables of the optimization problem correspond to each vector dimension. The first feasible food locations (solutions) are weighed by using an objective function to estimate which food location is an acceptable solution (nectaramount). Once the values from an objective function are obtained, the candidate solution $\mathbf{x}_i^k$ is modified through the ABC operators (honeybee types), where each food location $\mathbf{x}_i^k$ produces a new food source $\mathbf{t}_i$ and is calculated as below:

$$\mathbf{t}_i = \mathbf{x}_i^k + \phi(\mathbf{x}_i^k - \mathbf{x}_r^k), i, r \in (1, 2, \dots N); \tag{7.1}$$

where $\mathbf{x}_i^k$ is a food location randomly chosen, which obeys the following inequality $r \neq i$. ϕ is a random factor between $[-1, 1]$. After a new element $\mathbf{t}_i$ is produced, a probability correlated with the quality (fitness value) of the solution fit($\mathbf{t}_i$) is determined. With the Eq. (7.2) the fitness value is designated to the candidate solution $\mathbf{x}_i^k$.

$$\text{fit}(\mathbf{x}_i^k) = \begin{cases} \frac{1}{1+f(\mathbf{x}_i^k)} & \text{if } f(\mathbf{x}_i^k) \geq 0 \\ 1 + \left| f(\mathbf{x}_i^k) \right| & \text{if } f(\mathbf{x}_i^k) < 0 \end{cases}; \tag{7.2}$$

where $f(\cdot)$ is the objective function of the problem to be treated. after the fitness values are estimated, a greedy selection is employed between $\mathbf{t}_i$ and $\mathbf{x}_i^k$. If fit($\mathbf{t}_i$) is better than fit($\mathbf{x}_i^k$), the candidate solution $\mathbf{x}_i^k$ is replaced by $\mathbf{t}_i$; otherwise, $\mathbf{x}_i^k$ remains.

7.2.2 Differential Evolution (DE)

The DE is a technique developed by Storn and Price [29] based on a parallel direct search, the feasible solutions are represented as a vector and are described as follow:

$$\mathbf{x}_i^t = (\mathbf{x}_{1,i}^t, \mathbf{x}_{2,i}^t, \dots, \mathbf{x}_{d,i}^t), i = 1, 2, \dots n; \tag{7.3}$$

where $\mathbf{x}_i^t$ is the i-th element at the iteration t. The conventional DE technique has three principal steps: (a) *crossover*, (b) *mutation*, and (c) *selection* which are detailed below:

(A) *Crossover.* This step is controlled by a crossover rate $C_r \in [0, 1]$. The C_r determines the modification of some elements of the population and it is employed on each dimension of the solution. The crossover uses a uniformly distributed random number $r_i \in [0, 1]$ for determining the j-th component $\mathbf{v}_i$ which is manipulated as follows:

$$u_{j,i}^{t+1} = \begin{cases} \mathbf{v}_{j,i} & r_i \leq C_r \\ \mathbf{x}_{j,i}^t & \text{otherwise} \end{cases}, j = 1, 2, \dots, d \quad i = 1, 2, \dots, n; \tag{7.4}$$

(B) *Mutation.* This process is used on a given vector $\mathbf{x}_i^t$ at the t iteration. After three different elements of the population are randomly selected: $\mathbf{x}_p$, $\mathbf{x}_q$ and $\mathbf{x}_r$, a new candidate solution is generated by the following equation:

$$\mathbf{v}_i^{t+1} = \mathbf{x}_p^t + F \cdot (\mathbf{x}_q^t - \mathbf{x}_r^t); \tag{7.5}$$

(C) *Selection.* The selection process is basically the same that is used in any optimization process, the quality of the new solution is evaluated in the fitness function, if the new solution is better than the actual, then it is replaced. to determine the selection criteria, we can consider the following expression:

$$\mathbf{x}_i^{t+1} = \begin{cases} \mathbf{u}_i^{t+1} & \text{if } f(\mathbf{u}_i^{t+1}) \le f(\mathbf{x}_i^t) \\ \mathbf{x}_i^t & \text{otherwise} \end{cases} ; \tag{7.6}$$

7.2.3 Harmony Search (HS)

The HS was introduced by Geem [30], this technique is inspired by music. It was based on the perception that the meaning of music is to search for a perfect state of harmony by improvisation. The analogy in the optimization process is to search for the best value of a certain problem. When a musician improvises, he follows three rules for obtaining a good melody: plays a popular piece of music from his memory, plays something alike to a known piece, and composes a new one, or just plays arbitrary notes. In the HS, those rules become in three essential points; (a) *Harmony Memory (HM) consideration*, (b) *pitch adjustment*, and (c) *initialize a random solution.*

(a) *Harmony Memory (HM) Consideration.* This step will guarantee that the best harmonies will be transferred over to the new harmony memory. To handle the HM effectively it is designated a parameter that selects the elements of HM. This parameter is the Harmony Memory Consideration Rate ($HMCR$) and is described as $HMCR \in [0, 1]$. If this value is high, almost all the harmonies are employed in HM, and then other harmonies are not considered in the exploration stage. If the rate of the $HMCR$ is low, just a few best harmonies are chosen, and it may converge too slowly.

(b) *Pitch Adjustment.* Each element achieved by memory consideration is additionally examined to define whether it should be pitch-adjusted. For this operation, the Pitch-Adjusting Rate (PAR) is defined to select the frequency of the adjustment and the Bandwidth factor (BW) for controlling the local search nearby the chosen elements of the HM. Therefore, the pitch balancing decision is computed as follows:

$$x_{new} = \begin{cases} x_{new} \leftarrow x_{new} \pm \text{rand}(0,1) \cdot BW & \text{with probability } PAR \\ x_{new} & \text{with probability } (1 - PAR) \end{cases} ; \tag{7.7}$$

Pitch adjustment is responsible for producing new potential harmonies by slightly adjusting original variable positions. This process can be considered similar to the mutation method in evolutionary algorithms. Hence, the decision variable is either modified by a random number between 0 and BW or left unchanged.

(c) *Initialize a random solution.* This step tries to enhance the diversity of the population. As is specified in (a) if a new solution is not considered from the harmony memory it must be generated a new one in the search space. This process is defined as follows:

$$x_{new} = \begin{cases} x_i \in \{x_1, x_2, \ldots, x_{HMS}\} & \text{with probability } HMCR \\ l + (u - l \cdot \text{rand}(0, 1) & \text{with probability } (1\text{-}HMCR) \end{cases} ; \quad (7.8)$$

where *HMS* represents the size of the HM, x_i is an element taken from the HM, and *l* and *u* are the lower and upper bounds in the search space. Finally, rand is a random value uniformly distributed between 0 and 1.

7.2.4 Gravitational Search Algorithm (GSA)

Rashedi proposed the GSA in 2009, this evolutionary algorithm is based on the gravitational laws. In the GSA technique each mass (solution) is represented as a *d*-dimensional vector; $\mathbf{x}_i(t) = (x_i^1, \ldots, x_i^d)$; $(i \in 1, \ldots, N)$ at a time *t*, the acting force from a mass *i* to a mass *j* of the *h* variable is described as below:

$$F_{ij}^h(t) = G(t) \cdot \frac{Mp_i(t) \cdot Ma_j(t)}{R_{ij}(t) + \varepsilon} \cdot (x_j^h(t) - x_i^h(t)); \quad (7.9)$$

where Ma_j is the active gravitational mass associated to solution *j*, Mp_i expresses the passive gravitational mass of solution *i*, $G(t)$ is the gravitational constant at time *t*, ε is a small constant, and R_{ij} is the Euclidian distance between the *i*-th and *j*-th solutions. The $G(t)$ function is regulated through the evolution process to regulate the balance between exploration and exploitation by the adjustment of the attraction forces between particles. The total force acting over a certain solution *i* is described as follow:

$$F_i^h(t) = \sum_{j=1, j \neq i}^{N} F_{ij}^h(t); \quad (7.10)$$

Next, the acceleration of the candidate solution at time *t* is estimated as follows:

$$a_i^h(t) = \frac{F_i^h(t)}{Mn_i(t)}; \quad (7.11)$$

where Mn_i describes the inertial mass of the candidate solution *i*. The new position of each solution *i* is determined as below:

$$x_i^h(t + 1) = x_i^h(t) + v_i^h(t + 1)$$

$$v_i^h(t+1) = rand(\cdot) \cdot v_i^h(t) + a_i^h(t); \tag{7.12}$$

The gravitational and inertia masses of each solution are estimated in terms of their quality. Hence, the gravitational and inertia masses are updated by the following expression:

$$Ma_i = Mp_i = M_{ii} = M_i \tag{7.13}$$

$$m_i(t) = \frac{f(\mathbf{x}_i(t)) - \text{worst}(t)}{\text{best}(t) - \text{worst}(t)} \tag{7.14}$$

$$M_i(t) = \frac{m_i(t)}{\sum_{j=1}^{N} m_j(t)}, \tag{7.15}$$

where $f(\cdot)$ represents the objective function, whose result determines the quality of solutions.

7.2.5 Particle Swarm Optimization (PSO)

Probably the PSO is the most popular optimization method. It was introduced in 1995 by Kennedy [37] and it was inspired by the behavior of bird flocking. This technique employs a group of particles of N elements $\mathbf{p}^k(\{\mathbf{p}_1^k, \mathbf{p}_2^k, \ldots, \mathbf{p}_N^k\})$ as potential solutions, which are evolved from an initial position ($k = 0$) until reaching a certain stop criterion (Normally a total number of iterations). ($k = gen$) Each element $\mathbf{p}_i^k(i \in [1, \ldots, N])$ represents a d-dimensional vector $\{\mathbf{p}_{i,1}^k, \mathbf{p}_{i,2}^k, \ldots, \mathbf{p}_{i,d}^k\}$ where each decision variable of the optimization problem corresponds to a dimension. After the particle's random initialization within the search space, each individual $\mathbf{p}_i^k$ is evaluated by the use of a fitness function $f(\mathbf{p}_i^k)$, the best fitness value obtained during the evolution process is collected in a vector $\mathbf{g}(g_1, g_2, \ldots, g_d)$ with their corresponding positions $\mathbf{p}_i^*(\mathbf{p}_{i,1}^*, \mathbf{p}_{i,2}^*, \ldots, \mathbf{p}_{i,d}^*)$.

For practical proposes, the method used in the study was implemented by Paenke et al. in [38] here the PSO calculate the new position for each element as follows:

$$v_{i,j}^{k+1} = \omega \cdot v_{i,j}^k + c_1 \cdot r_1 \cdot (p_{i,j}^* - p_{i,j}^k) + c_2 \cdot r_2 \cdot (g_j - p_{i,j}^k),$$
$$p_{i,j}^{k+1} = p_{i,j}^k + v_{i,j}^{k+1} \tag{7.16}$$

where ω is an inertia weight, c_2 and c_1 regulates the updated and the current velocity, and are positive coefficients that control the acceleration velocity of each element towards the positions $\mathbf{g}$ and $\mathbf{p}_i^k$, respectively. r_1 and r_2 are uniformly distributed random numbers within the interval $[0, 1]$.

7.2.6 Cuckoo Search (CS) Technique

The CS method is one of the most popular nature-inspired techniques which has been introduced by Patwardhan et al. [39]. The CS algorithm is based on the family dependency of some cuckoo species. Moreover, this technique is improved by the method called Lévy flights [40], rather than by simple random walks. From the computational implementation, the CS generates a population $\mathbf{E}^k$ ($\{\mathbf{e}_1^k, \mathbf{e}_2^k, \ldots, \mathbf{e}_N^k\}$) of N eggs (elements) that evolves from the initial point ($k = 0$) to a total number of generations ($k = 2 \cdot gen$). Each egg $\mathbf{e}_i^k$ ($i \in [1, \ldots, N]$) describes a d-dimensional vector $\{e_{i,1}^k, e_{i,2}^k, \ldots, e_{i,d}^k\}$ where each dimension is a decision variable of the problem. The quality of each solution $\mathbf{e}_i^k$ (candidate solution) is evaluated by using an objective function $f\left(\mathbf{e}_i^k\right)$ whose result describes the fitness value of $\mathbf{e}_i^k$. Three main operators describe the optimization process of CS: (A) Lévy flight, (B) the Replace of Nests operator for creating new solutions, and (C) the selection procedure.

(A) The Lévy flight

$$\mathbf{s}_i = \frac{\mathbf{u}}{|\mathbf{v}|^{1/\beta}}, \tag{7.17}$$

where $\mathbf{u}$ ($\{u_1, \ldots, u_d\}$) and $\mathbf{v}$ ($\{v_1, \ldots, v_d\}$) are n-dimensional vectors and $\beta = 3/2$. Every element of $\mathbf{u}$ and $\mathbf{v}$ are determined by the following distributions:

$$u \sim N(0, \sigma_u^2), \quad v \sim N(0, \sigma_v^2),$$

$$\sigma_u = \left(\frac{\Gamma(1+\beta) \cdot \sin(\pi \cdot \beta/2)}{\Gamma((1+\beta)/2) \cdot \beta \cdot 2^{(\beta-1)/2}} \right)^{1/\beta}, \quad \sigma_v = 1, \tag{7.18}$$

where $\Gamma(\cdot)$ denotes the Gamma distribution. After $\mathbf{s}_i$ has been determined, the essential change of position $\mathbf{c}_i$ is calculated as follows:

$$\mathbf{c}_i = 0.01 \cdot \mathbf{s}_i \oplus (\mathbf{e}_i^k - \mathbf{e}^{best}), \tag{7.19}$$

where $\oplus$ expresses the entry-wise multiplications, while $\mathbf{e}^{best}$ is the best solution (egg) observed in terms of fitness value. Lastly, the new solution is $\mathbf{e}_i^{k+1}$ is computed by employing:

$$\mathbf{e}_i^{k+1} = \mathbf{e}_i^k + \mathbf{c}_i \tag{7.20}$$

(B) Replace some nests by constructing new solutions

This operator generates a set of elements that are probabilistically chosen and replaced with a new one. Each element $\mathbf{e}_i^k$ ($i \in [1, \ldots, N]$) can be picked with a probability $p_a \in [0, 1]$. To perform this action, a random number r_1 is produced within the range $[0, 1]$. If r_1 is lower than p_a, the element $\mathbf{e}_i^k$ is chosen and adjusted according to Eq. 7.5; otherwise $\mathbf{e}_i^k$ remains with no change. This operation can be described by the following model:

$$
\mathbf{e}_i^{k+1} = \begin{cases} \mathbf{e}_i^k + \mathrm{rand} \cdot (\mathbf{e}_j^k - \mathbf{e}_h^k) & \text{with probability } p_a \\ \mathbf{e}_i^k & \text{with probability } (1 - p_a) \end{cases} \tag{7.21}
$$

where *rand* is a random number while j and h are random integers from 1 to N.

(C) Elitist Selection

After generating $\mathbf{e}_i^{k+1}$ by using the operator A or B, it must be compared with its latest value $\mathbf{e}_i^k$. If the quality of $\mathbf{e}_i^{k+1}$ is better than $\mathbf{e}_i^k$, then $\mathbf{e}_i^{k+1}$ is taken as the final solution; otherwise, $\mathbf{e}_i^k$ remains. This method can be resumed as follows:

$$
\mathbf{e}_i^{k+1} = \begin{cases} \mathbf{e}_i^{k+1} & \text{if } f(\mathbf{e}_i^{k+1}) < f(\mathbf{e}_i^k) \\ \mathbf{e}_i^k & \text{otherwise} \end{cases} \tag{7.22}
$$

The selection strategy indicates that just high-quality eggs (best solutions near to the global optimal) which are the most similar to the host bird's eggs have the chance to improve (next generation) and grow mature cuckoos.

7.2.7 Differential Search Algorithm (DSA)

The Differential search algorithm was proposed by Civicioglu [33]. This technique imitates the Brownian-like random-walk move employed by an organism to migrate. Computationally, the DSA generates a population of individuals $\mathbf{X}^k(\mathbf{x}_1^k, \mathbf{x}_2^k, \ldots, \mathbf{x}_N^k)$ (artificial organisms) which are randomly initialized into the search space. Then, a *stopover* vector is produced characterized by the Brownian-like random-walks for each individual of the population, this stopover is described below:

$$
\mathbf{s}_{i,N} = \mathbf{X}_{i,N} + A \cdot (\mathbf{X}_{ri,N} - \mathbf{X}_{i,N}), \tag{7.23}
$$

where $ri \in [1, NP]$ is a random number in the population range and $ri \neq i$. A is a factor that controls the position variations of the elements. For the search strategy, the stopover position is computed as below:

$$
\mathbf{s}'_{i,j} = \begin{cases} \mathbf{s}_{i,j} & \text{if } r_{i,j} = 0, \\ \mathbf{X}_{i,j} & \text{if } r_{i,j} = 1, \end{cases} \tag{7.24}
$$

where $j = [1, .., d]$ and $r_{i,j}$ can be 0 or 1. When a candidate solution is selected, each element is evaluated by a cost function $f(\cdot)$ to estimate their quality, then a selection criterion is applied which is expressed as follows:

$$\mathbf{X}_i^{k+1} = \begin{cases} \mathbf{s}_i^k \text{ if } f\left(\mathbf{s}_i'\right) \leq f\left(\mathbf{X}_i^k\right), \\ \mathbf{X}_i^k \text{ if } f\left(\mathbf{s}_i'\right) > f\left(\mathbf{X}_i^k\right), \end{cases} \tag{7.25}$$

7.2.8 Crow Search Algorithm (CSA)

The crow search algorithm (CSA) was introduced by Askarzadeh [34], inspired by the intelligent behavior of crows. The CSA employs a population of $\mathbf{C}^k\left(\{\mathbf{c}_1^k, \mathbf{c}_2^k, .., \mathbf{c}_N^k\}\right)$ N elements (crows), where each element represents a d-dimensional vector $\left(\{\mathbf{c}_{i,1}^k, \mathbf{c}_{i,2}^k, .., \mathbf{c}_{i,m}^k\}\right)$. The search strategy of CSA can be described in two steps: The first step is when a crow is aware which means that is being followed by another crow; the second step is when is not aware. The steps of each crow are defined by a probability element AP_i^k. The new candidate solution is determined as below:

$$\mathbf{c}_i^{k+1} = \begin{cases} \mathbf{c}_i^k + r_i \times fl \times \left(\mathbf{m}_j^k - \mathbf{c}_i^k\right) \ r_j \geq AP_i^k, \\ \text{random position} \qquad\qquad \text{otherwise}, \end{cases} \tag{7.26}$$

where r_i and r_j are random numbers within the interval $[0, 1]$, fl is a parameter that regulates the flight length. $\mathbf{m}_j^k$ is the memory of crow j of where is saved the best solution at iteration k.

7.2.9 Covariant Matrix Adaptation with Evolution Strategy (CMA-ES)

The CMA-ES [35, 36] is an optimization technique introduced by Hansen based on the covariant matrix of the problem data. The CMA-ES applies a population of $\mathbf{X}^k\left(\{\mathbf{x}_1^k, \mathbf{x}_2^k, \ldots, \mathbf{x}_N^k\}\right)$ N elements that are randomly initialized. The selection of λ individuals of the next generation is determined as follows:

$$\mathbf{x}_N^{k+1} \sim N\left(\mathbf{x}_w^k, \sigma^{k^2} C^k\right), \tag{7.27}$$

where $N(\mu, C)$ is a normally distributed random vector with a mean μ and a covariance matrix C. The next weighted mean $\mathbf{x}_w^k$ chosen as the best interval is calculated as follows:

$$\mathbf{x}_w^k = \sum_{i=1}^{\mu} w_i \mathbf{x}_i^k, \tag{7.28}$$

where $\sum_{i=1}^{\mu} w_i = 1$. To modify the parameters, the CMA-ES applies two different arrangements, on the covariance matrix C^k and the global step size σ^k. For the covariance matrix adjustment case, an evolving path P_c^{k+1} is employed, which depends on the parent's division with $\mathbf{x}_w^k$ and the recombination points $\mathbf{x}_w^{k+1}$ as is shown below:

$$p_c^{k+1} = (1 - c_c)P_c^k + H_c^{k+1}\sqrt{c_c(2 - c_c)}\frac{\sqrt{\mu_{eff}}}{\sigma^k}(\mathbf{x}_w^{k+1} - \mathbf{x}_w^k), \tag{7.29}$$

$$C_c^{k+1} = (1 - c_{cov})C^k c_{cov}\frac{1}{\mu_{cov}}p_c^{k+1}\left(p_c^{k+1}\right)^T$$

$$+ c_{cov}\left(1 - \frac{1}{\mu_{cov}}\right)\sum_{i=1}^{\mu}\frac{w_i}{\sigma(k)}(\mathbf{x}_i^{k+1} - \mathbf{x}_i^k)(\mathbf{x}_i^{k+1} - \mathbf{x}_i^k), \tag{7.30}$$

$$H_\sigma^{k+1}\begin{cases} 1 & \frac{p_\sigma^{k+1}}{1-(1-c_\sigma)^{2(k+1)}} < \left(1.5 + \frac{1}{n-0.5}\right)E(\|N(0, I)\|), \\ 0 & \text{otherwise,} \end{cases} \tag{7.31}$$

where $\mu_{eff} = \left(\sum_{i=1}^{\mu} w_i\right)^2 \Big/ \sum_{i=1}^{\mu} w_i$ is the active variance range and $c_{cov} \approx \min\left(1, 2\mu_{eff}/n^2\right)$ is the learning rate. For the global step-size adjustment, a parallel path is used for modifying σ^k, this method is illustrated below:

$$p_\sigma^{k+1} = (1 - c_\sigma)p_\sigma^k + \sqrt{c_\sigma(2 - c_\sigma)}B^k D^{k^{-1}} B^{k^T} \times \frac{\mu_{eff}}{\sigma^k} \times (\mathbf{x}_w^{k+1} - \mathbf{x}_w^k), \tag{7.32}$$

where B^k is the orthogonal matrix and D^k is a diagonal matrix. The adjustment of global step size for the next generation is produced by the following equation:

$$\sigma^{k+1} = \sigma^k \exp\left(\frac{c_\sigma}{d_\sigma}\left(\frac{\|p_\sigma^{k+1}\|}{E(\|N(0, I)\|)} - 1\right)\right), \tag{7.33}$$

where $E(\|N(0, I)\|) = \sqrt{2}\Gamma(n + 1/2)\Big/\Gamma(n/2) \approx \sqrt{n}(1 - 1/4n + 1/21n^2)$ is the length of p_σ.

7.3 Solar Cells Modeling Process

Solar cells are the most popular and increasingly renewable energy source, under such circumstances, the correct modeling is an important task. Different alternatives for solar cells modeling have been introduced in the literature, such as employing neural networks [41–43]. Nevertheless, the most common modeling techniques are the equivalent circuits [5–7], which are described below.

Fig. 7.1 Single diode model

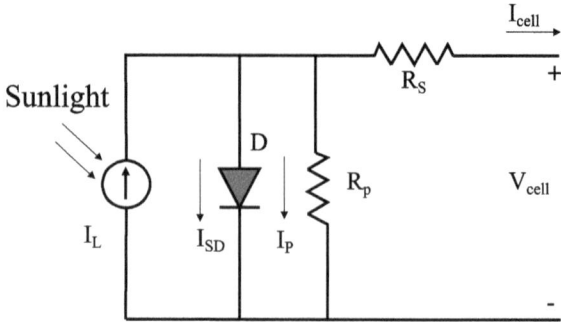

Single diode model (SDM)

The single diode equivalent model is the simpler and most accepted model for the description of solar cell's behavior. This representation employs a diode connected in parallel with the current source, which is shown in Fig. 7.1. Considering the circuit theorems, the total current of the model is determined as follows:

$$I_{cell} = I_L - I_{SD}\left\{\exp\left[\frac{q(V_{cell} + I_{cell}R_S)}{nkT}\right] - 1\right\} - \frac{V_{cell} + I_{cell}R_S}{R_p}, \quad (7.34)$$

where k is the Boltzmann constant, I_{SD} is the diffusion current, q is the electron charge, R_p and R_S are the parallel and serial resistances, and V_{cell} is the terminal voltage. The parameters that determinate the accuracy of the solar cell for the single diode model is given by five parameters; R_S, R_p, I_L, I_{SD} and n.

The double diode model is a different option to describe the solar cell characterization, where alternatively of one diode utilizes two in a parallel array as is presented in Fig. 7.2. The total current of two-diode model is calculated as:

$$I_{cell} = I_L - I_{D1} - I_{D2} - I_p, \quad (7.35)$$

where the diodes and leakage currents are determined as:

$$I_{D1} = I_{SD1}\left[\exp\left(\frac{q(V_{cell} + I_{cell}R_s)}{n_1kT}\right) - 1\right], \quad (7.36)$$

$$I_{D2} = I_{SD2}\left[\exp\left(\frac{q(V_{cell} + I_{cell}R_s)}{n_2kT}\right) - 1\right], \quad (7.37)$$

$$I_p = \frac{V_{cell} + I_{cell}R_s}{R_p}, \quad (7.38)$$

In this model the elements that must be accurately estimated are R_S, R_P, I_L, I_{SD1}, I_{SD2}, n_1 and n_2 which determinate the performance of the model.

Fig. 7.2 Double diode model

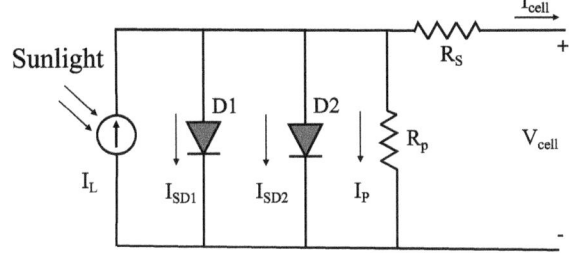

Three diode model (TDM)

The three-diode model is an alternative description of the solar cell models in which a third diode in parallel is included instead of two diodes, where a parameter n_3 must be taking into account in addition to the components of the two diodes as is shown in Fig. 7.3. Similarly, to the two-diode model, the total current is calculated as:

$$I_{cell} = I_L - I_{D1} - I_{D2} - I_{D3} - I_p, \tag{7.39}$$

where

$$I_{D1} = I_{SD1}\left[\exp\left(\frac{q(V_{cell} + I_{cell}R_{so}(1 + KI))}{n_1 kT}\right) - 1\right], \tag{7.40}$$

$$I_{D2} = I_{SD2}\left[\exp\left(\frac{q(V_{cell} + I_{cell}R_{so}(1 + KI))}{n_2 kT}\right) - 1\right], \tag{7.41}$$

$$I_{D3} = I_{SD3}\left[\exp\left(\frac{q(V_{cell} + I_{cell}R_{so}(1 + KI))}{n_3 kT}\right) - 1\right], \tag{7.42}$$

$$I_p = \frac{V_{cell} + I_{cell}R_{so}(1 + KI)}{R_p}, \tag{7.43}$$

In the three diode model, the element R_s was substituted by $R_{so}(1 + KI)$ to determine the difference of R_s with I_{cell}. Where I_{cell} is the load current and K is a

Fig. 7.3 Three diode model

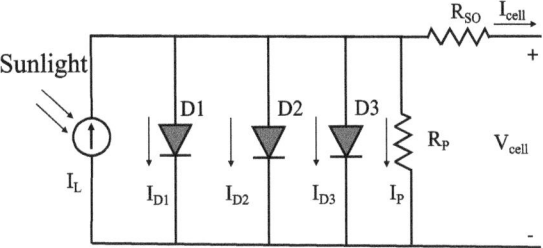

parameter that must be estimated as the rest of the parameters, for this, the parameters to be determined are R_{so}, R_p, I_L, I_{D1}, I_{D2}, I_{D3}, n_1, n_2, n_3 and K.

The array of solar cells can be expressed as modules [44, 45], which basically are the solar cells connected in parallel and serial. When the cells are configured in serial mode the voltage increases N_s times, in the case of the parallel configuration only the current increases in N_p times. Under such considerations, the output of a module of $N_s \times N_p$ cells is computed as follows:

$$I_m = N_p I_{cell}, \tag{7.44}$$

$$V_m = N_s V_{cell}, \tag{7.45}$$

$$R_{sm} = \frac{N_s}{N_p} R_s, \quad R_{pm} = \frac{N_s}{N_p} R_p, \tag{7.46}$$

Solar cells parameter identification as an optimization problem

The parameter estimation of solar cells can be handled as an optimization problem. The main objective is the accurate approximation between the true model and the equivalent circuit model. In the experimental results, the optimization techniques are assessed by a cost function to determine the quality of the solutions. To determine the solar cell parameters, the Eqs. (7.23), (7.24), and (7.28) are rewritten to reflex the variation of the experimental data as below:

$$f_{SDM}(V_{cell}, I_{cell}, \mathbf{x}) = I_{cell} - I_L + I_D + I_p, \tag{7.47}$$

$$f_{DDM}(V_{cell}, I_{cell}, \mathbf{x}) = I_{cell} - I_L + I_{D1} + I_{D2} + I_p, \tag{7.48}$$

$$f_{TDM}(V_{cell}, I_{cell}, \mathbf{x}) = I_{cell} - I_L + I_{D1} + I_{D2} + I_{D3} + I_p, \tag{7.49}$$

All equivalent models describe the parameters to be determined, single diode model (SDM) $\mathbf{x} = [R_s, R_p, I_L, I_{SD}, n]$, double diode model (DDM) $\mathbf{x} = [R_s, R_p, I_L, I_{SD1}, I_{SD2}, n_1, n_2]$, and three diode model (TDM) $\mathbf{x} = [R_{so}, R_p, I_L, I_{SD1}, I_{SD2}, I_{SD3}, n_1, n_2, n_3, K]$. The quality measure employed for the quantification of the solutions is the root mean square error (RMSE) which is described as:

$$RMSE = \sqrt{\frac{1}{N} \sum_{i=1}^{N} (f_i(V_{cell}, I_{cell}, \mathbf{x}))^2}, \tag{7.50}$$

Through the optimization process, the parameters are modified to minimize the cost function until a stop criterion is reached. Since the data acquisition is in irregular environmental conditions, the objective function implies noisy characteristics and multimodal error surface [46], which becomes the optimization process in a complex task [47].

7.4 Experimental Results

To perform the experimental results, the C60 mono-crystalline solar cell (SUNPOWER) was employed for the single and double diode models, for all cases, the D6P Multi-crystalline Photovoltaic Cell (DelSolar) was selected, due to the three diode model cells limitations [48], a KC200GT Photovoltaic Solar Module (Shell Solar) was incorporated just for the two first models. The solar cells parameters boundaries for each model are displayed in Table 7.1. The number of parameters for the Single Diode Model is five, in the Double Diode Model is 7 and finally for the Three Diode Model is 10, for the Three Diode Model 3 parameters are fixed [49]. I_{SC} is the short circuit current and R_s was substituted with $R_{so}(1 + KI)$ to find the variation for R_s concerning to I. The data acquisition was performed regarding the following operation conditions; (1000 W/m^2), (800 W/m^2), (500 W/m^2) and (300 W/m^2) at $T = 25\,^{\circ}\text{C}$. For the C60 Mono-crystalline two different sun irradiations are analyzed; and at, the RMSE average and standard deviation obtained from each technique in the SDM, and DD cases are performed after 40 individual executions and reported from Tables 7.2, 7.3, 7.4 and 7.5 respectively, where the CMA-ES determines the best accuracy in all approaches such as in the minimum value of the RSME as in the Average. For all case the number of $Search\ Agents = 50$ and the number of $Iterations = 2000$ in each execution.

In the Mono-crystalline C60 solar cell the best values of RMSE (minimum, maximum, and average) were collected by the CMA-ES except for the standard deviation in which the best values were given by the CS in the single diode case at

Table 7.1 Solar cells parameters range for the three diode models

Single diode model			Double diode model			Three diode model		
Parameter	Lower	Upper	Parameter	Lower	Upper	Parameter	Lower	Upper
$R_s\,(\Omega)$	0	0.5	$R_s\,(\Omega)$	0	0.5	$R_{so}\,(\Omega)$	0	1
$R_p\,(\Omega)$	0	200	$R_p\,(\Omega)$	0	100	$R_p\,(\Omega)$	0	100
$I_L\,(A)$	I_{SC}		$I_L\,(A)$	I_{SC}		$I_L\,(A)$	I_{SC}	
$I_{SD}\,(\mu A)$	0	1	$I_{SD1}\,(\mu A)$	0	1	$I_{SD1}\,(\mu A)$	0	1
n_1	1	2	$I_{SD1}\,(\mu A)$	0	1	$I_{SD2}\,(\mu A)$	0	1
–	–	–	n_1	0	2/3	$I_{SD3}\,(\mu A)$	0	1
–	–	–	n_2	0	2/3	n_1	1	
–	–	–	–	–	–	n_2	2	
–	–	–	–	–	–	n_3	0	3
–	–	–	–	–	–	K	0	1

Table 7.2 Solar cells (mono-crystalline C60) parameters determination, mean and standard deviation for SDM and DDM at 1000 W/m^2

SDM

Parameters	ABC	CSA	CS	DE	DSA	GSA	HS	PSO	CMA-ES
RS (Ω)	0.005065	0.006535176	0.00168	0.00466	0.0051363	0.00542089	0.00920948	0.0034763	0.00662421
Rp (Ω)	183.463	114.252273	199.3241	199.4603	152.35831	79.83066	164.621448	39.481739	200
IL (A)	6.240779	6.21459738	6.546991	6.23187	6.2295673	3.53701901	6.41504223	6.2343094	6.21432845
ID (A)	2.08E−06	2.79016E−07	6.27E−08	3.3E−06	2.197E−06	2.4077E−06	5.3421E−06	3.595E−06	2.41E−07
n	1.697168	1.496370914	1.345742	1.7507	1.7033435	1.87810036	1.82172571	1.7523015	1.48357172
Min RMSE	0.010921	0.009935587	0.267794	0.010551	0.0099479	0.03315632	0.01324157	0.0113077	**0.00992256**
Max RMSE	0.014243	0.01156138	0.268449	0.012094	0.0115475	0.20423651	0.07939054	0.0369322	**0.0104022**
Average RMSE	0.011869	0.010522777	0.268275	0.011191	0.0106624	0.1058941	0.02830248	0.0169241	**0.00995462**
Std	0.00066	0.000461744	0.000122	0.000316	0.0004252	0.04410207	0.01343111	0.0047521	**0.00010586**

DDM

Parameters	ABC	CSA	CS	DE	DSA	GSA	HS	PSO	CMA-ES
RS (Ω)	0.005678	0.006503674	0.002098	0.006623	0.006368	4.6383E−05	0.00585544	0.0058646	6.62E−03
Rp (Ω)	104.5934	191.2923658	188.4067	135.4169	161.64551	89.7438772	23.2336847	70.063423	2.00E+02
IL (A)	6.226711	6.214465293	6.54696	6.216998	6.2162283	3.29376151	6.09719654	6.2251079	6.21432845
ID1 (A)	2.52E−06	2.98848E−09	6.48E−07	1.29E−06	2.551E−07	3.5163E−06	8.3191E−07	8.32E−07	2.41E−07
ID2 (A)	1.01E−06	2.86929E−07	2.73E−08	1.91E−07	1.984E−07	9.9952E−07	9.2192E−07	5.353E−07	4.16E−16
n1	2.996087	1.548379914	1.936987	2.13222	1.4991135	1.96725788	1.7084817	1.8053035	1.48357169

(continued)

Table 7.2 (continued)

DDM									
Parameters	ABC	CSA	CS	DE	DSA	GSA	HS	PSO	CMA-ES
n2	1.619807	1.499321137	1.290596	1.466079	1.6776963	1.90130136	1.66715067	1.5634685	1.48324541
Min RMSE	0.010186	0.009922907	0.267669	0.009963	0.0099278	0.01742442	0.01243733	0.0104161	**0.00992256**
Max RMSE	0.011947	0.021592727	0.26857	0.03142	0.0309824	0.05321138	0.04337821	0.0137103	**0.01167399**
Average RMSE	0.010878	0.018137755	0.268162	0.014355	0.0128385	0.03266309	0.01869163	0.0115566	**0.00994816**
Std	0.000528	0.000230786	**0.000189**	0.000341	0.0002764	0.00833028	0.00624268	0.0008043	0.00124354

Table 7.3 Solar cells (mono-crystalline C60) parameters estimation, mean and standard deviation for SDM and DDM at 800 W/m^2

SDM

Parameters	ABC	CSA	CS	DE	DSA	GSA	HS	PSO	CMA-ES
RS (Ω)	0.000488	0.000804375	2E-05	0.000608	0.0009882	0.00513349	0.00031329	0.0206406	0.00163057
Rp (Ω)	120.7382	98.41301429	192.0449	199.9999	131.57376	105.754974	77.0013758	66.393912	24.8933861
IL (A)	4.882213	4.881234281	5.130919	4.878856	4.8774982	3.76620098	4.70079256	4.8472653	4.88245696
ID (A)	2.15E-06	1.80815E-06	2.96E-09	2.12E-06	1.464E-06	2.9701E-06	3.8413E-06	6.636E-06	6.87E-07
n	1.728899	1.709113752	1.170159	1.727269	1.684991	1.85447388	1.80017479	1.8717672	1.60471698
Min RMSE	0.004947	0.004868946	0.490195	0.004944	0.004869	0.01796734	0.00803614	0.0059932	**0.00486895**
Max RMSE	0.006511	0.00633468	0.490546	0.005489	0.0051846	0.16861586	0.0824437	0.0320081	**0.0050412**
Average RMSE	0.005311	0.005094385	0.490393	0.005165	0.0049925	0.08699095	0.02791164	0.0169787	**0.00487734**
Std	0.000262	0.000331104	7.47E-05	0.000131	7.706E-05	0.03347491	0.01654392	0.0070283	**3.0309E-05**

DDM

Parameters	ABC	CSA	CS	DE	DSA	GSA	HS	PSO	CMA-ES
RS (Ω)	0.001232	0.001737993	4.04E-05	0.001437	0.0012232	0.00013346	0.00077359	0.0005271	0.00163057
Rp (Ω)	119.4226	74.60410516	192.1913	31.65655	92.170396	94.7860962	38.135595	54.42826	24.8933897
IL (A)	4.87627	4.876536455	5.130963	4.880101	4.8766252	2.94761087	4.87654863	4.8827055	4.88245696
ID1 (A)	1.33E-06	1.72118E-06	6.3E-10	1.25E-06	7.495E-07	1.1397E-06	5.4769E-06	1.97E-06	4.12E-20
ID2 (A)	0.004192	2.67912E-07	1.94E-08	6.06E-07	4.097E-07	9.9984E-07	2.4922E-07	3.501E-07	6.87E-07
n1	1.674555	1.920364455	1.090282	2.045905	1.6671119	1.9393706	1.89673872	1.7292267	1.96919786

(continued)

Table 7.3 (continued)

DDM

Parameters	ABC	CSA	CS	DE	DSA	GSA	HS	PSO	CMA-ES
n2	51.2116	1.533094915	1.999292	1.598242	1.6469724	1.81085015	1.65114007	1.812426	1.60471696
Min RMSE	0.004884	0.004870705	0.489958	0.004879	0.00487	0.01021258	0.00666149	0.0049378	**0.00486895**
Max RMSE	0.005309	0.005923922	0.490594	0.005059	0.0053881	0.03878291	0.02295133	0.0077401	**0.00505602**
Average RMSE	0.005044	0.004957579	0.490274	0.004934	0.0050934	0.02353969	0.01157486	0.0057249	**0.00491157**
Std	0.000144	0.000171857	0.000156	3.71E−05	0.0001448	0.00730923	0.00383631	0.0008088	**3.6698E−05**

Table 7.4 Solar cells (mono-crystalline C60) parameters estimation, mean and standard deviation for SDM and (DDM at 500 W/m^2)

SDM

Parameters	ABC	CSA	CS	DE	DSA	GSA	HS	PSO	CMA-ES
RS (Ω)	0.0016	0.003883448	1.17E−05	0.003187	0.0039048	0.00821181	0.00642602	0.4181931	0.00441815
Rp (Ω)	163.4687	199.973256	198.2291	200	130	115.937296	109.512364	168.0123	48.9235971
IL (A)	3.038953	3.037506852	3.364666	3.039108	3.0385235	1.59703402	3.09297001	3.0363817	3.04073716
ID (A)	2.33E−06	7.66902E−07	4.89E−11	1.16E−06	7.455E−07	8.4011E−07	7.4442E−06	1.78E−06	4.93E−07
n	1.769269	1.640749248	1.006255	1.685994	1.6376616	1.94585248	1.93745184	1.7383762	1.59453154
Min RMSE	0.004097	0.0039777713	0.775675	0.004006	0.0039789	0.00621321	0.00573558	0.0043468	**0.00397771**
Max RMSE	0.004518	0.00485107	0.775887	0.004348	0.0043833	0.05531805	0.03164257	0.0079782	**0.00404263**
Average RMSE	0.004288	0.00408547	0.775706	0.004139	0.0040739	0.03202631	0.01165965	0.0055138	**0.00398091**
Std	9.52E−05	0.000195093	3.73E−05	8.61E−05	9.076E−05	0.01185082	0.00541076	0.0009191	**1.2258E−05**

DDM

Parameters	ABC	CSA	CS	DE	DSA	GSA	HS	PSO	CMA-ES
RS (Ω)	0.004034	0.004177475	7.24E−05	0.003613	0.0041682	0.00056852	6.1811E−05	0.0009983	0.00441815
Rp (Ω)	77.26164	53.46679376	199.3801	190.1412	76.437153	83.6005286	105.243007	193.65696	48.9235809
IL (A)	3.045323	3.03984558	3.364619	3.03873	3.0393953	1.84818457	3.0452358	3.0414082	3.04073717
ID1 (A)	6.91E−07	3.52681E−06	6.22E−11	8.18E−07	7.987E−10	1.8939E−06	5.2676E−06	3.468E−06	4.93E−07
ID2 (A)	0.000382	6.38146E−08	1.01E−10	5.59E−07	5.927E−07	9.9958E−07	6.8983E−07	9.414E−07	2.05E−20
n1	1.630061	1.995274001	1.63944	1.950576	1.7433395	1.98517022	1.88922847	1.9159403	1.59453153

(continued)

Table 7.4 (continued)

DDM

Parameters	ABC	CSA	CS	DE	DSA	GSA	HS	PSO	CMA-ES
n2	30.69521	1.439759222	1.03605	1.617409	1.6134584	1.91104148	1.99442602	1.7420063	1.9973227
Min RMSE	0.003986	0.003977746	0.775707	0.003984	0.003978	0.00519559	0.00485494	0.0040001	**0.00397771**
Max RMSE	0.004303	0.00445058	0.776099	0.004639	0.0041926	0.01890252	0.01283665	0.0046005	**0.00407281**
Average RMSE	0.004106	0.004025824	0.775866	0.004214	0.0040236	0.01149578	0.00692778	0.00416	**0.00402082**
Std	0.000113	9.35314E−05	9.82E−05	**2.21E−05**	5.435E−05	0.0032242	0.00151241	0.0001555	0.00016829

Table 7.5 Solar cells (mono-crystalline C60) parameters estimation, mean and standard deviation for SDM and DDM at 300 W/m^2

SDM

Parameters	ABC	CSA	CS	DE	DSA	GSA	HS	PSO	CMA-ES
RS (Ω)	0.022856	0.024935901	1.28E−07	0.021591	0.019553	0.0057405	0.01079812	0.0157213	0.03784345
Rp (Ω)	89.60072	96.99719894	199.9309	200	196.57055	95.1381775	117.815101	198.31323	20.3660429
IL (A)	1.815495	1.799703082	2.16471	1.804586	1.8048883	0.647616	1.81085624	1.8076504	1.8066546
ID (A)	6.67E−07	1.66103E−07	1.75E−11	6.07E−07	1.356E−06	2.2364E−08	4.8415E−06	4.25E−06	4.88E−11
n	1.680349	1.537152887	1.000012	1.669162	1.7645843	1.97885237	1.93524352	1.9192075	1.02519908
Min RMSE	0.005007	0.004163427	0.972854	0.004777	0.0041294	0.00669648	0.00603242	0.0050737	**0.00324028**
Max RMSE	0.00604	0.005120555	0.972854	0.005612	0.0055624	0.02650582	0.0148986	0.0069059	**0.00484652**
Average RMSE	0.005533	0.004597845	0.972854	0.005207	0.0049094	0.01435318	0.00831713	0.0059562	**0.00364383**
Std	0.000167	0.000266064	**7.89E−08**	0.000162	0.0003	0.00500556	0.00195439	0.0003223	0.00060098

DDM

Parameters	ABC	CSA	CS	DE	DSA	GSA	HS	PSO	CMA-ES
RS (Ω)	0.019759	0.009998557	1.11E−06	0.019881	0.0099999	0.00294815	0.00572879	0.0049867	0.03453452
Rp (Ω)	108.4338	190.951882	198.7353	199.9866	199.49798	89.6325187	29.6722844	60.11981	1.28E+02
IL (A)	1.808202	1.801598482	2.164701	1.806197	1.8026873	1.64108059	1.79601671	1.7960872	1.80E+00
ID1 (A)	1.4E−06	5.36334E−06	1.3E−09	1.12E−06	6.436E−06	2.6725E−06	2.3832E−06	5.283E−06	2.87E−08
ID2 (A)	9.51E−07	9.9852E−07	1.77E−11	6.06E−07	9.032E−07	9.8344E−07	7.7925E−07	5.803E−07	1.12E−09
n1	1.767764	1.999925106	1.941844	1.744111	1.9999988	1.97590137	1.8494644	1.968805	1.85070386

(continued)

Table 7.5 (continued)

DDM

Parameters	ABC	CSA	CS	DE	DSA	GSA	HS	PSO	CMA-ES
n2	4.789302	1.892108949	1.0004	2.260392	1.999976	1.81370144	1.96612708	1.9296981	1.18E+00
Min RMSE	0.004836	0.00683517	0.972855	0.004646	0.0068351	0.0076357	0.00749499	0.0069709	**0.00349494**
Max RMSE	0.00596	0.007264676	0.972864	0.005748	0.0068425	0.015570402	0.01230902	0.0075616	**0.00573986**
Average RMSE	0.005606	0.006895311	0.972859	0.005281	0.0068354	0.01074707	0.00871806	0.0071482	**0.00489288**
std	0.000299	8.98654E−05	2.38E−06	0.000227	**1.212E−06**	0.00152859	0.00095817	0.0001411	0.0005975

(1000 W/m^2), in the double diode the DE at (800 W/m^2), and for single and double diode respectively at (300 W/m^2) the CS and DSA. The parameter setting for the experimental test are determined according to their references in which have shown by experimentation the best possible combination for solar cell parameter estimation, these parameters are described below:

1. ABC: the parameter limit is set at 100 [17].
2. DE: $F = 0.4$ and $C_r = 0.4$ [50].
3. HS: According to [22], $HMRC = 0.95$, and $PAR = 0.3$.
4. GSA: the parameters are set at $G_0 = 100$ and $\alpha = 20$ [31].
5. PSO: Agreeing to [51], $c_1 = 0.5$, $c_2 = 2.5$, the weight factor decreases from 0.9 to 0.4.
6. CS: The $p_a = 0.25$ in concordance with [24].
7. DSA: $p_1 = 0.3$ and $p_2 = 0.3$ regarding to [33].
8. CSA: The $AP = 0.1$, and $fl = 2$ [34].
9. CMA-ES: The technique used is described in [35, 36].

For the Multi-crystalline solar cell D6P case, two different solar irradiations are considered; (1000 W/m^2) and (500 W/m^2) at $T = 25$ °C for the three different models, SDM, DDM, and TDM, where the optimization techniques defined the best solar cell parameters which are presented in Tables 7.6 and 7.7 respectively with the mean and standard deviation to verify their performance. The CMA-ES outperform the results of the other algorithms for the case of SDM, DDM, and TDM.

For the Mono-crystalline D6P, the CMA-ES determine the most accurate average and minimum RMSE for the three cases, but in the case of standard deviation, the ABC in the single diode at (1000 W/m^2), DE in the three diode model at (1000 W/m^2) and CS in the single and double diode at (500 W/m^2) achieved the best results. To define the capabilities of the adopted optimization algorithms, we employ the Multi-crystalline module KC200GT for the Single Diode and Double Diode Models (1000 W/m^2), (800 W/m^2), (600 W/m^2), (400 W/m^2) and (200 W/m^2) where the mean, average, and standard deviation are detailed from Tables 7.8, 7.9, 7.10, 7.11 and 7.12 respectively. In the Multi-crystalline module, the CMA-ES presents competitive results determining the minimum average in most of the cases, but the CS and DE can find good solutions with important accuracy.

For the solar cell module, the CMA-ES achieved the most suitable result for almost all cases, except for the standard deviation in various tests as the CS in the single diode model (800 W/m^2) and (600 W/m^2). The DSA determines the best standard deviation for the single and double diode model (400 W/m^2) and (200 W/m^2).

To validate the experimental results, we statistically analyze the data collected for the optimization techniques. For this, we employed the Wilcoxon analysis [50, 51], which estimates the difference between two related techniques. In this test, we consider 5% of the significance level of the Average RSME between the compared techniques, where the p-value set for the pair-wise similarity between techniques is 0.05. After analyzing specific considerations between techniques employed for the comparative study, we examined the follows groups: CMA-ES versus ABC,

Table 7.6 Solar cells (mono-crystalline D6P) parameters determination, mean and standard deviation for SDM, DDM, and TDM at 1000 W/m²

SDM

Parameters	ABC	CSA	CS	DE	DSA	GSA	HS	PSO	CMA-ES
RS (Ω)	4.93E−03	0.00550083	0.006453	0.005662	0.005988	0.005038	0.001861	0.4709873	0.47098731
Rp (Ω)	7.50E+01	100	99.45801	100	99.99615	66.07779	80.98159	57.768198	57.7681979
IL (A)	8.31E+00	8.30252074	8.299179	8.300678	8.294515	6.897731	8.23392	8.2149683	8.21496834
ID (A)	1.57E−06	5.34E−07	6.53E−08	3.51E−07	1.64E−07	1E−05	7.31E−06	8.249E−06	8.2489E−06
n	1.606934	1.50376702	1.335087	1.466456	1.404038	1.90333	1.776386	1.7966804	1.79668045
Min RMSE	0.023204	0.02320388	0.015645	0.019481	0.019044	0.034982	0.034982	0.0260146	**0.01552954**
Max RMSE	0.025538	0.02553847	0.023851	0.023045	0.024382	0.216904	0.216904	0.0447331	**0.02003157**
Average RMSE	0.024574	0.02457448	0.019163	0.020916	0.021132	0.108073	0.108073	0.0311407	**0.01892337**
Std	**0.00069**	0.00068966	0.002722	0.000794	0.001218	0.047229	0.047229	0.0036418	0.00082729

DDM

Parameters	ABC	CSA	CS	DE	DSA	GSA	HS	PSO	CMA-ES
RS (Ω)	5.70E−03	0.00555699	0.006699	0.006064	0.006557	0.002284	0.003773	0.0041813	0.00418127
Rp (Ω)	6.71E+01	100	89.09031	100	99.96804	35.31513	55.90306	79.052877	79.0528766
IL (A)	8.29E+00	8.30146867	8.281023	8.296526	8.288297	7.172523	8.287227	8.3190704	8.31907043
ID1 (A)	3.82E−06	3.92E−07	3.55E−08	1.57E−07	1.71E−10	3.94E−06	4.96E−06	4.837E−06	4.8375E−06
ID2 (A)	2.48E−07	4.30E−07	2.22E−11	1.8E−07	4.63E−08	1E−06	9.89E−07	1.198E−08	1.198E−08
n1	5.71E+00	1.88991907	1.293362	23.35675	1.176865	1.86705	1.770647	1.7332564	1.73325637
n2	1.438037	1.48667571	1.389624	1.412048	1.313473	1.62115	1.712209	1.8487766	1.84877659
Min RMSE	0.02146	0.02146031	0.017369	0.019226	0.017316	0.029422	0.029422	0.0215297	**0.01707178**
Max RMSE	0.025601	0.02560146	0.027343	0.023371	0.02305	0.06036	0.06036	0.0281581	**0.02145989**

(continued)

Table 7.6 (continued)

DDM

Parameters	ABC	CSA	CS	DE	DSA	GSA	HS	PSO	CMA-ES
Average RMSE	0.023643	0.02364294	0.024494	0.021652	0.020183	0.039861	0.039861	0.0252211	**0.01940024**
Std	0.001206	0.0012056	0.002065	0.001091	0.001305	0.00789	0.00789	0.0016887	**0.00098192**

TDM

Parameters	ABC	CSA	CS	DE	DSA	GSA	HS	PSO	CMA-ES
I_{D1} (A)	1.21E−10	1.177E−10	1.21E−10	8.59E−26	1.21E−10	1.23E−10	1.24E−10	1.249E−08	1.2486E−08
I_{D2} (A)	8.23E−11	2.1463E−07	2.13E−13	3.31E−05	3.82E−21	9.20E−13	7.97E−07	2.312E−07	2.3123E−07
I_{D3} (A)	4.92E−04	9.9377E−06	1.2E−15	2.52E−19	3.14E−14	2.33E−10	8.69E−06	7.016E−06	7.0156E−06
$n3$	8.80E+01	2.41406574	0.863338	2.993156	98.52522	2.996657	2.97427	2.6504631	2.65046313
R_{so} (Ω)	8.02E−03	0.00738269	0.008045	0.002745	0.008045	0.008008	0.005064	0.0063443	0.00634432
K	6.74E−04	0.01435204	9.35E−09	1.74E−16	1.91E−15	0.004806	0.016666	0.0474956	0.04749557
R_{sh} (Ω)	1.84E+00	2.12835817	1.819064	4.077476	1.818718	99.99369	36.85799	69.683667	69.6836673
Min RMSE	0.020393	0.02039341	0.020388	0.020393	0.019618	0.023934	0.062702	0.0262095	**0.01949509**
Max RMSE	0.020471	0.02047111	0.020393	0.020393	0.03713	0.0333	39.09412	0.0550369	**0.02038819**
Average RMSE	0.020404	0.02040431	0.020393	0.020393	0.025195	0.0535	9.684074	0.0400163	**0.02010918**
Std	1.62E−05	1.6213E−05	9.3E−07	**5.66E−09**	0.00665	0.000455	8.682488	0.0052987	0.00028239

Table 7.7 Solar cells (mono-crystalline D6P) parameters determination, mean and standard deviation for SDM, DDM, and TDM at 500 W/m^2

SDM

Parameters	ABC	CSA	CS	DE	DSA	GSA	HS	PSO	CMA-ES
RS (Ω)	0.004157	0.01159328	2E−09	0.005909	0.00681	0.001227	0.000849	0.0029151	0.00749966
Rp (Ω)	69.17906	100	99.99913	100	99.99975	35.77021	94.14903	43.577466	100
IL (A)	4.156335	4.13161166	4.6471	4.142898	4.141783	2.647298	4.202613	4.1544498	4.140932
ID (A)	1.49E−06	2.27E−10	7.22E−11	3.12E−07	1.51E−07	7.05E−06	9.55E−06	3.263E−06	7.25E−08
n	1.586101	1	1.000001	1.435838	1.376114	1.933243	1.807051	1.6732462	1.31973729
Min RMSE	0.017136	0.01713641	0.567476	0.016359	0.015404	0.018751	0.018751	0.017759	**0.01416909**
Max RMSE	0.018008	0.01800797	0.567477	**0.017199**	0.017287	0.076983	0.076983	0.0188433	0.01740387
Average RMSE	0.017607	0.01760732	0.567476	0.016775	0.016542	0.043495	0.043495	0.0183816	**0.01611262**
STD	0.000187	0.00018661	**8.51E−08**	0.00021	0.000376	0.013287	0.013287	0.0002893	0.00111236

DDM

Parameters	ABC	CSA	CS	DE	DSA	GSA	HS	PSO	CMA-ES
RS (Ω)	6.51E−03	0.00711864	7.29E−05	0.005624	0.006064	0.00253	0.004874	0.0055739	0.00557385
Rp (Ω)	7.98E+01	100	99.91123	100	61.38597	49.47878	82.65916	82.816173	82.816173
IL (A)	4.16E+00	4.14223473	4.646989	4.144588	4.144383	4.158399	4.146255	4.1468956	4.14689559
ID1 (A)	8.07E−07	1.19E−07	8.61E−11	1.59E−06	2.08E−07	4.32E−06	9.28E−07	7.139E−06	7.1389E−06
ID2 (A)	1.87E−07	9.57E−08	9.06E−10	1.32E−07	9.99E−07	9.94E−07	8.06E−07	3.587E−07	3.5869E−07
n1	3.66E+01	1.35906364	1.007153	1.800635	1.404134	1.81406	1.980265	1.5796227	1.5796227
n2	1.392767	1.77294891	1.950849	1.380061	1.999837	1.603871	1.528525	1.4499956	1.44999562
Min RMSE	0.015887	0.01588696	0.567477	0.016084	0.01645	0.017171	0.017171	0.0165929	**0.01519835**
Max RMSE	0.017786	0.01778611	0.567613	**0.017333**	0.018129	0.019554	0.019554	0.0174407	0.01735582

(continued)

Table 7.7 (continued)

DDM

Parameters	ABC	CSA	CS	DE	DSA	GSA	HS	PSO	CMA-ES
Average RMSE	0.017235	0.01723534	0.567503	0.016818	0.017597	0.018417	0.018417	0.017041	**0.01624427**
std	0.000408	0.000408	**2.78E−05**	0.000279	0.000373	0.000565	0.000565	0.0002042	0.00048383

TDM

Parameters	ABC	CSA	CS	DE	DSA	GSA	HS	PSO	CMA-ES
ID1 (A)	2.10E−10	2.1006E−10	1.58E−13	1.55E−10	6.72E−11	1.21E−10	1.65E−10	4.144E−08	4.1445E−08
ID2 (A)	3.24E−15	2.1954E−07	2.93E−10	6.89E−21	6.02E−19	5.47E−07	6.31E−07	4.874E−07	4.8742E−07
ID3 (A)	4.83E−10	3.4658E−11	7.69E−14	1.02E−15	3.12E−13	1.19E−14	9.28E−06	4.896E−06	4.8958E−06
n3	9.32E+01	1.78535981	0.784333	0.684267	0.792991	0.99053	1.945619	2.1120764	2.11207644
Rso (Ω)	1.03E−02	0.00999985	0.001608	0.011271	0.01	0.007961	0.002745	0.0071011	0.00710112
K	4.35E−11	2.4762E−06	0.002556	4.17E−15	0.039095	0.003907	0.098125	0.0547249	0.05472488
Rsh (Ω)	7.10E−01	0.70491482	99.97324	0.707655	0.69295	67.37059	0.632458	70.815981	70.8159813
Min RMSE	0.042226	0.04222615	0.567153	0.042226	0.041832	0.043415	0.089162	0.0439661	**0.04059946**
Max RMSE	0.042227	0.04222669	0.567277	0.042226	0.042156	1.6245	5.652148	0.0800799	**0.04197422**
Average RMSE	0.042226	0.04222618	0.5672	0.042226	0.04195	0.2893	1.757516	0.0592006	**0.04142937**
Std	1.03E−07	1.0326E−07	3.12E−05	**9.81E−18**	0.000115	0.000455	1.565291	0.0102721	0.00034344

Table 7.8 Solar cells Module (multi-crystalline KC200GT) parameters determination, mean and standard deviation for SDM and DDM at 1000 W/m^2

SDM

Parameters	ABC	CSA	CS	DE	DSA	GSA	HS	PSO	CMA-ES
RS (Ω)	0.00819725	0.008598412	0.009702	0.008254	0.0085027	0.00532522	0.006372	0.0086455	0.00838946
Rp (Ω)	14.2746543	287.7649217	278.3281	300	274.24082	162.599163	145.730272	70.236383	500
IL (A)	8.45231967	8.401473158	8.395708	8.405986	8.4081122	5.51397782	8.1667034	8.4517976	8.40253114
ID (A)	2.1304E−06	8.80807E−07	1.02E−07	1.24E−06	1.218E−06	4.8699E−06	9.1479E−06	5.298E−06	1.23E−06
n	1.63367968	1.542895531	1.361663	1.575057	1.5744236	1.83695064	1.81751613	1.7462203	1.57520904
Min RMSE	0.00619726	0.005158409	0.006143741	0.006055	0.0051727	0.01676816	0.0077667	0.0071031	**0.00506031**
Max RMSE	0.00819386	0.006852389	0.0006919949	0.006854	0.0066646	0.08468712	0.03851485	0.0182919	**0.00594658**
Average RMSE	0.00697368	0.006031249	0.006439478	0.00623	0.0059189	0.05137936	0.01626893	0.0110058	**0.00536039**
Std	0.00051988	0.000375625	0.000262026	0.000192	0.0003831	0.0164289	0.00766401	0.0026008	**0.00013732**

DDM

Parameters	ABC	CSA	CS	DE	DSA	GSA	HS	PSO	CMA-ES
RS (Ω)	0.01018216	0.010537688	0.010568	0.010349	0.0105334	0.00579965	0.0125366	0.4753595	0.01040301
Rp (Ω)	136.897439	133.9638333	264.3318	299.9873	299.97365	169.410993	245.068672	141.23624	197.337864
IL (A)	8.37444967	8.371571382	8.371119	8.371028	8.3703729	4.12627659	8.01575802	8.074969	8.37548908
ID1 (A)	6.439E−08	2.32597E−15	7.8E−09	1.36E−08	3.258E−13	9.9792E−09	4.8326E−09	2.546E−09	1.15E−10
ID2 (A)	2.2274E−08	8.47844E−09	2.55E−09	2.69E−08	8.692E−09	9.9249E−09	5.1778E−09	1.771E−09	1.25E−08
n1	8.73132142	1.351574739	1.194356	1.227365	1.1265378	1.6933772	1.17566595	1.5137775	1.24036008

(continued)

Table 7.8 (continued)

DDM

Parameters	ABC	CSA	CS	DE	DSA	GSA	HS	PSO	CMA-ES
n2	1.25726919	1.199160692	1.875205	1.747515	1.2006042	1.89036558	1.51451236	1.1106203	1.22215402
Min RMSE	0.00473273	0.004717362	0.004821636	0.004718	0.0047173	0.01310982	0.00566407	0.0061483	**0.00471767**
Max RMSE	0.00518495	0.004856829	0.004741333	0.004753	0.004756	0.09772879	0.06769827	0.0190041	**0.00472145**
Average RMSE	0.00487024	0.004737393	0.00482412	0.004728	0.0047213	0.05200618	0.0274881	0.0114809	**0.004719**
Std	0.00011032	2.80154E−05	8.61837E−07	8.93E−06	6.524E−06	0.01691836	0.01305747	0.0041682	**7.0567E−06**

Table 7.9 Solar cells Module (multi-crystalline KC200GT) parameters estimation, mean and standard deviation for SDM and DDM at 800 W/m^2

SDM

Parameters	ABC	CSA	CS	DE	DSA	GSA	HS	PSO	CMA-ES
RS (Ω)	0.00560748	7.82E−05	2.84E−05	5.75E−03	1.08E−03	0.00473677	1.76E−03	0.5313321	0.00392955
Rp (Ω)	156.472585	499.9948083	472.2883	500	499.99812	2.86E+02	115.897865	98.146888	500
IL (A)	6.74823457	5.13E+00	5.130815	6.745251	5.131	1.35389938	5.13099889	2.9315738	5.131
ID (A)	3.56E−06	5.16E−09	3.67E−07	2.53E−06	1.80E−08	7.58E−08	4.55E−06	3.48E−06	2.60E−10
n	1.6482632	1.187243554	1.441785	1.610137	1.2151826	1.84E+00	1.71599654	1.7138988	1.00003708
Min RMSE	0.00607673	0.153164765	0.153071491	0.006144	0.1530621	0.15859113	0.15484765	0.1549824	**0.00596914**
Max RMSE	0.00770129	0.154739618	0.153109032	0.006847	0.1532002	0.21342771	0.1611923	0.1601034	**0.00681995**
Average RMSE	0.0069878	0.153856882	0.153088145	0.006439	0.1531086	0.16944026	0.15669185	0.15735	**0.00637779**
Std	0.00037055	0.000400204	**8.81371E−06**	0.00023	3.457E−05	0.01004141	0.00143392	0.0013584	0.00015752

DDM

Parameters	ABC	CSA	CS	DE	DSA	GSA	HS	PSO	CMA-ES
RS (Ω)	0.00652112	0.002510291	0.004005	0.007576	0.0039691	0.00323222	8.39E−04	2.61E−01	0.00400321
Rp (Ω)	290.306499	471.3428232	499.4977	5.564204	5.00E+02	2.75E+02	2.55E+02	365.18762	500
IL (A)	6.72995842	5.130923625	5.131	6.775167	5.1309999	2.85E+00	5.13E+00	3.89E+00	5.131
ID1 (A)	7.78E−07	1.48E−10	4.82E−10	4.07E−08	2.58E−10	1.96E−10	9.69E−09	6.76E−09	2.57E−10
ID2 (A)	7.30E−07	1.71E−10	2.56E−10	6.39E−08	3.03E−11	9.92E−09	4.43E−09	5.24E−09	8.86E−21
n1	1.49260116	1.014655345	1.787907	1.347456	1.0001267	1.73017437	1.17802325	1.808445	1

(continued)

Table 7.9 (continued)

DDM

Parameters	ABC	CSA	CS	DE	DSA	GSA	HS	PSO	CMA-ES
n2	50.4069533	1.764172097	1.00E+00	1.317329	1.9897464	1.61841891	1.66892515	1.1611895	1.99938964
Min RMSE	0.00474219	0.153022239	0.152992432	0.004718	0.1529846	0.1550537	0.15346329	0.1534739	**0.00471636**
Max RMSE	0.00535054	0.153360344	0.153025772	0.004818	0.1530329	0.1682246	0.16383126	0.1604568	**0.00475133**
Average RMSE	0.00495841	0.153098018	0.15301046	0.00473	0.1530076	0.1619196	0.15583403	0.1556357	**0.00472412**
Std	0.00016899	6.08071E−05	9.66935E−06	1.58E−05	1.29E−05	0.00284984	0.00231309	0.0015596	**7.5167E−06**

Table 7.10 Solar cells Module (multi-crystalline KC200GT) parameters determination, mean and standard deviation for SDM and DDM at 600 W/m^2

SDM

Parameters	ABC	CSA	CS	DE	DSA	GSA	HS	PSO	CMA-ES
RS (Ω)	0.00579404	4.69E−06	0.000354	0.00471	1.99E−09	0.0239241	3.23E−04	0.4360552	9.29E−19
Rp (Ω)	103.380418	494.4540364	4.19E+02	11.75391	499.99974	2.75E+02	297.465108	2.40E+02	500
IL (A)	5.03219147	3.364709848	3.36464	5.064251	3.36471	1.52204555	3.36E+00	2.97E+00	3.36471
ID (A)	4.38E−07	5.72E−11	6.08E−08	1.25E−06	4.86E−10	1.76E−09	2.17E−06	2.79E−06	2.61E−10
n	1.43825761	1.00107902	1.320566	1.535283	1.029505	1.926606	1.68277169	1.76E+00	1
Min RMSE	0.00656172	0.240433034	0.240272335	0.006144	0.2402715	0.24175207	0.24147391	0.2415572	**0.00589603**
Max RMSE	0.00937572	0.241981687	0.240302873	0.007021	0.2404957	0.26679805	0.24479858	0.2442355	**0.00681995**
Average RMSE	0.00737129	0.241182992	0.240281109	0.006459	0.2403316	0.24944009	0.24224258	0.2426385	**0.00623948**
Std	0.00064136	0.000392438	**8.27765E−06**	0.000235	5.655E−05	0.00521166	0.00076707	0.0006874	0.00013652

DDM

Parameters	ABC	CSA	CS	DE	DSA	GSA	HS	PSO	CMA-ES
RS (Ω)	0.00663807	1.81E−06	7.46E−06	7.65E−03	5.30E−12	3.95E−03	1.67E−03	0.0156249	6.69E−18
Rp (Ω)	6.47586816	493.0062779	4.80E+02	3.829101	499.99997	2.51E+02	3.67E+02	2.41E+02	500
IL (A)	5.06785663	3.364709691	3.364706	5.10E+00	3.36471	2.08921992	3.36404429	8.69E−01	3.36471
ID1 (A)	1.11E−07	5.25E−11	3.02E−10	2.82E−10	2.61E−10	4.09E−10	6.27E−09	5.54E−09	2.61E−10
ID2 (A)	1.28E−08	5.64E−11	9.76E−09	1.59E−08	5.06E−15	4.18E−10	4.48E−10	4.12E−09	1.69E−20
n1	1.32627965	1.813680317	1.007273	1.287049	1	1.67217836	1.16332975	1.4373337	1

(continued)

Table 7.10 (continued)

DDM

Parameters	ABC	CSA	CS	DE	DSA	GSA	HS	PSO	CMA-ES
n2	16.9103492	1.000368025	1.772075	1.195659	1.9984145	1.58346052	1.61249948	1.1380601	1.98601177
Min RMSE	0.00474395	0.240326451	0.2402723	0.004719	0.2402715	0.24114319	0.24074788	0.2410345	**0.00471636**
Max RMSE	0.00527847	0.241718986	0.240276037	0.004751	0.2402715	0.24875306	0.24933665	0.2423547	**0.00474795**
Average RMSE	0.00494073	0.240753244	0.240274015	0.004727	0.2402715	0.24357299	0.24184863	0.2417584	**0.00472412**
Std	0.00013252	0.000424668	9.58666E−07	6.2E−06	**2.082E−09**	0.00155382	0.00139945	0.000263	7.8217E−06

Table 7.11 Solar cells module (multi-crystalline KC200GT) parameters determination, mean and standard deviation for SDM and DDM at 400 W/m^2

SDM

Parameters	ABC	CSA	CS	DE	DSA	GSA	HS	PSO	CMA-ES
RS (Ω)	0.0082898	8.73E−09	0.00013	0.008543	1.01E−09	8.84E−03	0.00262176	0.4076231	2.47E−16
Rp (Ω)	244.788126	499.9985195	5.00E+02	15.25072	499.99657	226.435541	364.481124	341.52765	5.00E+02
IL (A)	3.37145929	2.164709998	2.164709	3.388371	2.16E+00	1.47930568	2.16E+00	1.1096869	2.16E+00
ID (A)	2.24E−06	1.64E−11	1.04E−08	2.03E−06	4.69E−10	8.90E−09	9.91E−07	1.72E−06	2.43E−10
n	1.61640042	1.000000439	1.20E+00	1.606847	1.029216	1.69801672	1.63653468	1.6495691	1
Min RMSE	0.00645975	0.3009123307	0.300845399	0.006144	0.3008454	0.30158912	0.30100127	0.3010593	**0.00543773**
Max RMSE	0.01174902	0.302867207	0.300845464	0.007065	0.3008454	0.32471022	0.3035138	0.3040074	**0.00681995**
Average RMSE	0.00764025	0.301076997	0.300845412	0.006528	0.3008454	0.3082867	0.30152288	0.3023678	**0.00643948**
Std	0.00094938	0.000375184	1.30416E−08	0.000307	**1.585E−15**	0.0050065	0.00053007	0.0007353	0.00025252

DDM

Parameters	ABC	CSA	CS	DE	DSA	GSA	HS	PSO	CMA-ES
RS (Ω)	0.01188791	2.04E−07	6.97E−05	0.012755	1.98E−12	2.24E−02	1.25E−03	2.70E−01	8.77E−18
Rp (Ω)	16.8146454	5.00E+02	448.477	5.031745	5.00E+02	2.45E+02	3.64E+02	4.35E+02	500
IL (A)	3.37079874	2.16E+00	2.164703	3.41E+00	2.16E+00	2.16E+00	2.16E+00	9.38E−01	2.16471
ID1 (A)	1.67E−07	1.66E−11	2.50E−10	4.89E−09	1.74E−15	1.00E−08	9.39E−09	6.64E−09	1.71E−21
ID2 (A)	1.33E−09	3.20E−11	7.17E−09	4.18E−08	2.43E−10	9.98E−09	7.20E−09	1.86E−09	2.43E−10
n1	1.36986811	1.000009475	1.000271	2.454186	1.9889716	1.78140319	1.26138542	1.5581188	1.98855852

(continued)

Table 7.11 (continued)

DDM

Parameters	ABC	CSA	CS	DE	DSA	GSA	HS	PSO	CMA-ES
n2	310.20159	1.95998006	1.857724	1.266697	1	1.80164641	1.22777934	1.1065305	1
Min RMSE	0.00476202	0.300849656	0.300845428	0.004718	0.3008454	0.30106051	0.30097447	0.3009636	**0.00378198**
Max RMSE	0.00633534	0.301050402	0.300845888	0.004845	0.3008454	0.30564935	0.30159409	0.3014975	**0.00476034**
Average RMSE	0.00501231	0.300961924	0.30084558	0.004725	0.3008454	0.3023613	0.30114073	0.301172	**0.0045456**
Std	0.00026864	3.31744E−05	9.054E−08	8.71E−06	**7.976E−14**	0.00092921	0.0001444	0.0001434	8.7065E−06

Table 7.12 Solar cells module (multi-crystalline KC200GT) parameters determination, mean and standard deviation for SDM and DDM at 200 W/m²

SDM

Parameters	ABC	CSA	CS	DE	DSA	GSA	HS	PSO	CMA-ES
RS (Ω)	0.01141702	5.44E−08	1.61E−02	1.05E−02	0.0145769	0.01762537	0.0018854	1.79E−01	0.02447924
Rp (Ω)	147.191212	499.9912692	167.8188	201.0279	339.93184	2.21E+02	1.70E+02	429.52154	24.3117548
IL (A)	1.68087497	2.164709998	1.675382	1.676926	1.6757034	1.41175765	1.68040673	1.6786414	1.68007665
ID (A)	3.78E−07	1.65E−11	7.55E−08	6.07E−07	1.66E−07	9.70E−06	6.38E−06	2.40E−06	4.50E−10
n	1.43443606	1.000000995	1.300344	1.481847	1.3638056	1.91865929	1.75798573	1.6323506	1
Min RMSE	0.00640138	0.300942015	0.3008454	0.006144	0.3008454	0.30114476	0.30098639	0.3011497	**0.00606314**
Max RMSE	0.00965965	0.302323073	0.300845451	0.007015	0.3008454	0.32671827	0.30270375	0.3048593	**0.00681995**
Average RMSE	0.00744433	0.301011783	0.30084541	0.00653	0.3008454	0.30861766	0.30154735	0.3025211	**0.00643948**
Std	0.00077881	0.000243937	1.10173E−08	0.000198	**5.385E−16**	0.00578398	0.00043297	0.0008678	0.00043752

DDM

Parameters	ABC	CSA	CS	DE	DSA	GSA	HS	PSO	CMA-ES
RS (Ω)	0.02301243	6.55E−08	0.024463	0.022756	0.0240381	1.65E−02	0.02084871	2.29E−02	2.45E−02
Rp (Ω)	257.694441	499.1710598	23.34492	51.2072	3.62E+01	255.333055	402.097715	4.02E+02	24.3117565
IL (A)	1.67617312	2.164709912	1.680472	1.68E+00	1.68E+00	1.60E+00	1.66829732	1.6706903	1.68E+00
ID1 (A)	3.47E−09	1.00E−11	4.50E−10	1.66E−09	6.67E−10	4.74E−11	6.83E−09	1.84E−09	1.47E−21
ID2 (A)	1.08E−09	1.65E−11	2.71E−09	8.54E−10	2.93E−10	7.87E−09	7.08E−09	2.45E−09	4.50E−10
n1	6.73073562	1.92E+00	1.000026	1.062261	1.018037	1.64570227	1.19967205	1.0677641	1.86977567

(continued)

Table 7.12 (continued)

DDM

Parameters	ABC	CSA	CS	DE	DSA	GSA	HS	PSO	CMA-ES
n2	1.04000367	1.000039816	1.777667	2.290439	1.9428294	1.14876342	1.17225141	1.670724	1
Min RMSE	0.00472589	0.300855771	0.300845424	0.004718	0.3008454	0.30118528	0.30096118	0.3009839	**0.00471636**
Max RMSE	0.00535393	0.301022601	0.300845711	0.004851	0.3008454	0.30648166	0.30158047	0.3015689	**0.00474499**
Average RMSE	0.00494432	0.30095187	0.300845559	0.004725	0.3008454	0.30264489	0.30111403	0.3011762	**0.00472412**
Std	0.00016301	3.00989E−05	7.51683E−08	6.7E−06	**6.492E−14**	0.00130048	0.00014352	0.0001382	6.1847E−06

CMA-ES versus CSA, CMA-ES versus CS, CMA-ES versus DE, CMA-ES versus DSA, CMA-ES versus HS, and CMA-ES versus PSO, this, due to the performance presented in the experimental tests, where the CMA-ES showed better consistency for the solar cells model estimation. In the Wilcoxon analysis, the null hypothesis considers that there is no statistical difference between approaches, and as an alternative hypothesis that exists a significant difference between both approaches. On the other hand, if the number of elements in the test is large, the chance to fall in the error type 1 increases, for this, the significance value must be corrected applying the Bonferroni correction [52, 53]. Once we determine the new significance value (n-value), the result computed by Wilcoxon is compared with the n-value, if the result is lower, the null hypothesis is rejected, avoiding the error type 1. To simplify this analysis, in Table 7.13, the symbols ▲, ► and ▼ are adopted, where ▲ means that the technique examined performs better than compared technique, ▼ indicates that the technique determine worse results than the compared algorithm, and ► represents that there is no difference between techniques. The n-value defined by Bonferroni correction was calculated as $n = 0.00139$.

According to the statistical analysis, after determining the value of Bonferroni correction as is presented in Table 7.13, in the C60 Mono-crystalline, Mono-crystalline D6P solar cells, and the Multi-crystalline KC200GT, the CMA-ES outperforms (▲) the techniques adopted for the study in the three equivalent models (Average RMSE). Considering the typical statistical criteria, the standard deviation showed the most varying results, where the CMA-ES, DE, CS, and DSA performed similar conclusions. After analyzing Table 7.13, we observe that the CMA-ES can determine better solutions in most cases. These results exhibit that exists a statistical difference among the CMA-ES and the rest of the approaches used for this comparison [54].

In Fig. 7.4, the I-V features between the measured data and the estimated models determined by the CMA-ES are presented, in this case, two different conditions are examined; in condition A, the irradiation is set as 1000 W/m^2. For condition B, the irradiation is 500 W/m^2. The D6P100 Multi-crystalline solar cell was adopted for the graphical representation for the SDM, DDM, and TDM (Figs. 7.5 and 7.6).

Table 7.13 Statistical analysis for the SDM and DDM for the Mono-crystalline cell and the Multi-crystalline Module and SDM, DDM and TDM for the Multi-crystalline cell at different irradiation conditions after the Bonferroni correction

C60 mono-crystalline

IR		CMA-ES versus							
		ABC	CROW	CS	DE	DS	GSA	HS	PSO
1000	SDM	1.43E−14 ◄	1.43E−14 ◄	1.28E−14 ◄	1.28E−14 ◄	1.51E−13 ◄	1.33E−14 ◄	1.39E−14 ◄	1.43E−14 ◄
	DDM	1.01E−05 ◄	0.100873 ◄	1.53E−14 ◄	1.64E−14 ◄	5.2E−05 ◄	1.45E−14 ◄	1.14E−13 ◄	6.90E−08 ◄
800	SDM	1.66E−14 ◄	1.66E−14 ◄	1.30E−14 ◄	1.28E−14 ◄	1.03E−12 ◄	1.34E−14 ◄	1.58E−14 ◄	1.43E−14 ◄
	DDM	1.76E−05 ◄	0.176388 ◄	1.35E−14 ◄	1.51E−14 ◄	1.12E−07 ◄	1.48E−14 ◄	1.46E−14 ◄	7.21E−08 ◄
500	SDM	1.43E−14 ◄	1.43E−14 ◄	1.29E−14 ◄	1.29E−14 ◄	3.58E−13 ◄	1.42E−14 ◄	1.63E−14 ◄	1.43E−14 ◄
	DDM	1.29E−06 ◄	0.012867 ◄	1.24E−14 ◄	1.39E−14 ◄	1.74E−05 ◄	1.50E−14 ◄	1.45E−14 ◄	1.56E−08 ◄
300	SDM	1.44E−14 ◄	1.57E−08 ◄	1.44E−14 ◄	1.67E−12 ◄	2.22E−12 ◄	1.35E−13 ◄	1.54E−14 ◄	1.31E−14 ◄
	DDM	4E−09 ◄	1.44E−14 ◄	1.38E−14 ◄	1.39E−14 ◄	1.44E−14 ◄	1.28E−14 ◄	1.53E−14 ◄	1.36E−14 ◄

(continued)

Table 7.13 (continued)

D6P100 Multi-crystalline		CMA-ES versus							
IR		ABC	CROW	CS	DE	DS	GSA	HS	PSO
1000	SDM	2.81E−14 ◄	9.96E−06 ◄	1.99E−06 ◄	5.84E−04 ◄	1.74E−04 ◄	1.44E−14 ◄	1.44E−14 ◄	1.44E−14 ◄
	DDM	5.04E−04 ◄	1.28E−12 ◄	2.08E−12 ◄	1.56E−09 ◄	1.38E−11 ◄	1.44E−14 ◄	2.22E−12 ◄	2.22E−12 ◄
	TDD	8.66E−04 ◄	3.47E−04 ◄	6.91E−08 ◄	5.10E−05 ◄	7.98E−09 ◄	1.43E−14 ◄	2.92E−12 ◄	2.92E−12 ◄
500	SDM	4.08E−14 ◄	6.56E−04 ◄	1.24E−14 ◄	1.23E−04 ◄	1.59E−04 ◄	1.44E−14 ◄	1.44E−14 ◄	1.44E−14 ◄
	DDM	3.44E−05 ◄	7.91E−14 ◄	1.44E−14 ◄	1.69E−11 ◄	1.64E−13 ◄	6.73E−09 ◄	1.11E−12 ◄	1.11E−12 ◄
	TDDD	1.06E−14 ◄	1.06E−14 ◄	1.06E−14 ◄	1.06E−14 ◄	1.06E−14 ◄	1.06E−14 ◄	1.06E−14 ◄	1.06E−14 ◄

(continued)

Table 7.13 (continued)

Mod. kc200gt mono-crystalline

IR		CMA-ES versus							
		ABC	CROW	CS	DE	DS	GSA	HS	PSO
1000	SDM	2.08E−11 ◄	1.36E−14 ◄	2.36E−11 ◄	7.18E−04 ◄	6.55E−08 ◄	2.85E−14 ◄	1.28E−14 ◄	2.44E−14 ◄
	DDM	1.94E−14 ◄	1.63E−14 ◄	1.63E−14 ◄	2.79E−05 ◄	1.55E−06 ◄	1.51E−14 ◄	1.24E−14 ◄	2.44E−14 ◄
800	SDM	1.58E−11 ◄	1.35E−14 ◄	2.44E−11 ◄	4.05E−04 ◄	7.44E−08 ◄	2.75E−14 ◄	1.25E−14 ◄	2.54E−14 ◄
	DDM	2.81E−14 ◄	1.45E−14 ◄	1.44E−14 ◄	9.31E−05 ◄	2.14E−06 ◄	1.61E−14 ◄	1.22E−14 ◄	2.50E−14 ◄
600	SDM	6.35E−14 ◄	1.36E−14 ◄	2.44E−11 ◄	4.73E−04 ◄	5.44E−08 ◄	3.14E−14 ◄	1.30E−14 ◄	2.49E−14 ◄
	DDM	1.55E−14 ◄	1.44E−14 ◄	1.44E−14 ◄	1.13E−05 ◄	1.35E−06 ◄	1.71E−14 ◄	1.24E−14 ◄	2.44E−14 ◄
400	SDM	1.94E−12 ◄	1.40E−14 ◄	2.44E−11 ◄	2.71E−04 ◄	7.84E−08 ◄	2.94E−14 ◄	1.25E−14 ◄	2.44E−14 ◄
	DDM	1.44E−14 ◄	1.49E−14 ◄	1.44E−14 ◄	7.47E−05 ◄	1.51E−06 ◄	1.55E−14 ◄	1.23E−14 ◄	2.42E−14 ◄
200	SDM	8.1E−12 ◄	1.33E−14 ◄	2.44E−11 ◄	2.72E−04 ◄	6.74E−08 ◄	3.12E−14 ◄	1.21E−14 ◄	2.41E−14 ◄
	DDM	3.27E−14 ◄	1.41E−14 ◄	1.44E−14 ◄	2.13E−05 ◄	1.46E−06 ◄	1.89E−14 ◄	1.27E−14 ◄	2.49E−14 ◄

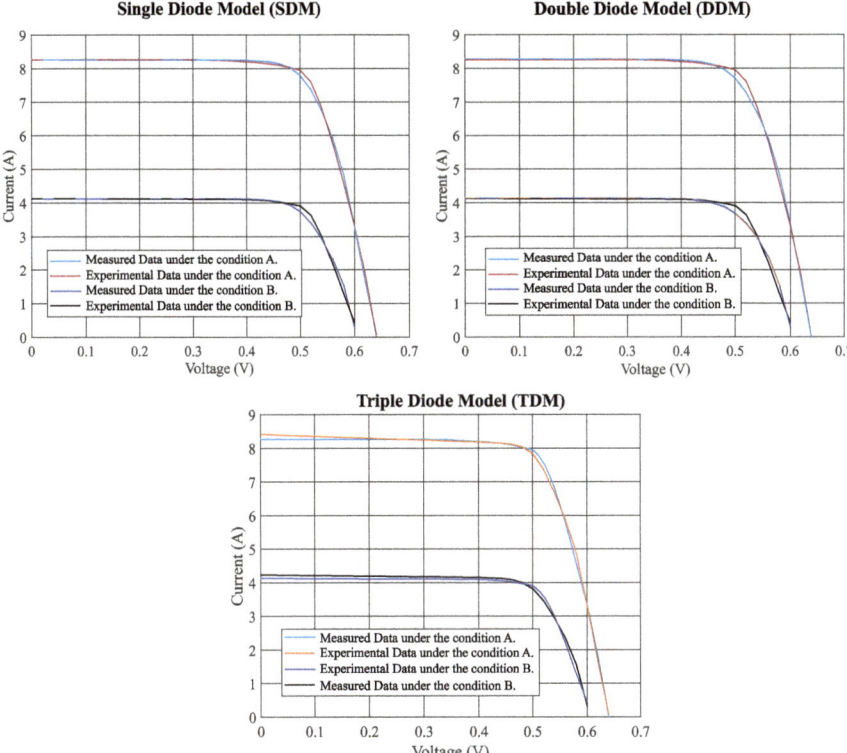

Fig. 7.4 I-V characteristic between the measured data and the approximate model computed by CMA-ES for the D6P100 Multi crystalline solar cell under two irradiation conditions: condition A and condition B

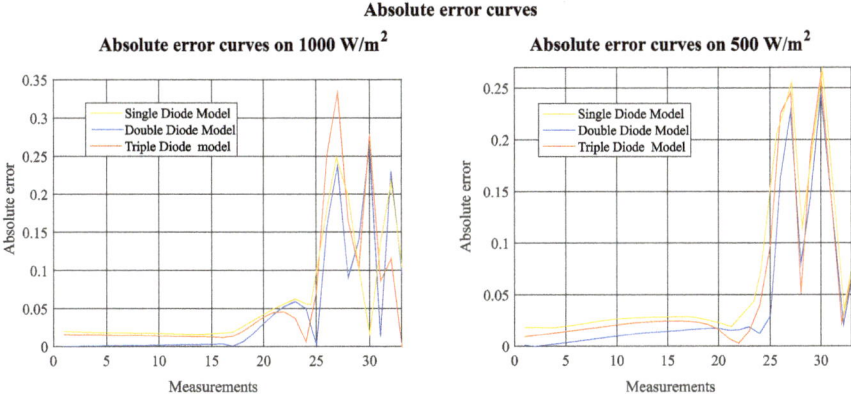

Fig. 7.5 Absolute error curves generated by the CMA-ES for the D6P100 Multi-crystalline solar cell under two irradiation conditions: condition A and condition B for the SDM, DDM and TDM

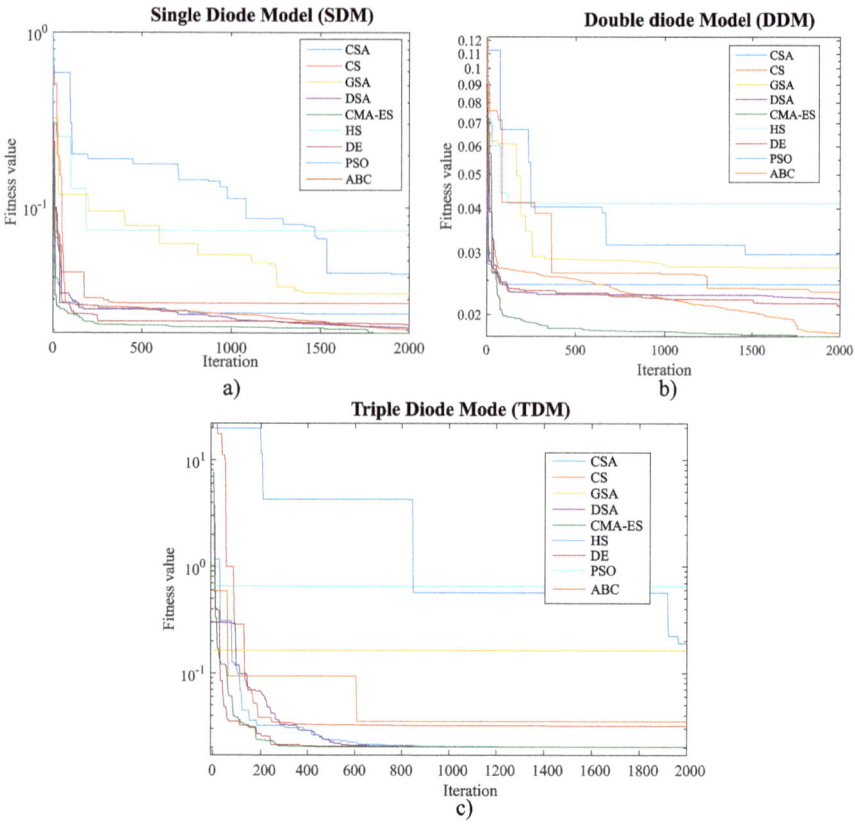

Fig. 7.6 Convergence evolution over iterations of the optimization techniques for the SDM (**a**), DDM (**b**) and TDM (**c**) employing the D6P100 multi-crystalline solar cell

7.5   Conclusions

In this chapter, comparative research for solar cells parameter determination is treated. In the study, several optimization techniques are employed, such as Artificial Bee Colony (ABC), Crow Search Algorithm (CSA), Cuckoo Search (CS), Differential Evolution (DE), Differential Search (DSA), Gravitational Search Algorithm (GSA), Harmony Search (HS), Particle Swarm Optimization (PSO) and Covariant Matrix Adaptation with Evolution Strategy (CMA-ES). This work was generated over three different equivalent solar cell models, Single Diode (SDM), Double Diode Model (DDM), and Three Diode Model (TDM) employing a Mono-crystalline solar cell for the SDM and DDM, a Multi-crystalline solar cell for the SDM, DDM, and TDM, and a solar cell module for the SDM and DDM.

Since the solar cell parameter determination is a complex task due to its dimensionality and multimodality, the correct comparative study of different approaches to determine which can perform the best results for solar cell parameter estimation.

After comparing the capabilities of the optimization techniques to estimate the best parameters for the three diode models, the CMA-ES showed the best performance regarding the rest algorithms adopted for the study for the minimum and average root mean square error (RMSE).

In terms of the standard deviation, the results were divided, highlighting the CMA-ES, CS, DE, and DSA. The experimental results of these techniques are due to the good valance between exploration and exploitations such as the operators employed for their search strategy. On the other hand, the DE registered a good convergence rate almost like the CMA-ES, the worst convergence rate is for the PSO and HS in which in several instances did not obtain good solutions. To statistically verify the experimental results, a non-parametric test known as Wilcoxon's test was employed as well as the usual statistical tests such as average, standard deviation, minimum and maximum values, where in most cases the CMA-ES achieved the best results. On the other hand, to avoid error type 1, a Bonferroni correction was chosen, where a new significance value was determined, verifying that the CMA-ES statistically outperforms the rest of the algorithms used for the study.

References

1. Shafiee S, Topal E (2009) When will fossil fuel reserves be diminished? Energy Policy 37:181–189. https://doi.org/10.1016/j.enpol.2008.08.016
2. Peer review of renewables 2017 global status report—REN21. http://www.ren21.net/peer-review-renewables-2017-global-status-report/. Accessed 21 Mar 2017
3. Town C (2012) CIE42 Proceedings, 16–18 July 2012, Cape Town, South Africa © 2012 CIE & SAIIE, 16–18
4. Quaschning V, Hanitsch R (1996) Numerical simulation of current-voltage characteristics of photovoltaic systems with shaded solar cells. Sol Energy 56:513–520. https://doi.org/10.1016/0038-092X(96)00006-0
5. Farivar G, Asaei B (2010) Photovoltaic module single diode model parameters extraction based on manufacturer datasheet parameters. PECon2010—2010 IEEE Int Conf Power Energy 929–934. https://doi.org/10.1109/PECON.2010.5697712
6. Sudhakar Babu T, Prasanth Ram J, Sangeetha K et al (2016) Parameter extraction of two diode solar PV model using Fireworks algorithm. Sol Energy 140:265–276. https://doi.org/10.1016/j.solener.2016.10.044
7. Khanna V, Das BK, Bisht D et al (2015) A three diode model for industrial solar cells and estimation of solar cell parameters using PSO algorithm. Renew Energy 78:105–113. https://doi.org/10.1016/j.renene.2014.12.072
8. Easwarakhanthan T, Bottin J, Bouhouch I, Boutrit C (1986) Nonlinear minimization algorithm for determining the solar cell parameters with microcomputers. Int J Sol Energy 4:1–12. https://doi.org/10.1080/01425918608909835
9. Ortiz-Conde A, García Sánchez FJ, Muci J (2006) New method to extract the model parameters of solar cells from the explicit analytic solutions of their illuminated I-V characteristics. Sol Energy Mater Sol Cells 90:352–361. https://doi.org/10.1016/j.solmat.2005.04.023
10. Jain A, Kapoor A (2004) Exact analytical solutions of the parameters of real solar cells using Lambert W-function. Sol Energy Mater Sol Cells 81:269–277. https://doi.org/10.1016/j.solmat.2003.11.018

11. Saleem H, Karmalkar S (2009) An analytical method to extract the physical parameters of a solar cell from four points on the illuminated $J - V$ curve. Electron Device Lett IEEE 30:349–352. https://doi.org/10.1109/LED.2009.2013882

12. Appelbaum J, Peled A (2014) Parameters extraction of solar cells—a comparative examination of three methods. Sol Energy Mater Sol Cells 122:164–173. https://doi.org/10.1016/j.solmat.2013.11.011

13. Zagrouba M, Sellami A, Bouaı M (2010) Identification of PV solar cells and modules parameters using the genetic algorithms: application to maximum power extraction. Sol Energy 84:860–866. https://doi.org/10.1016/j.solener.2010.02.012

14. Bastidas-Rodriguez JD, Petrone G, Ramos-Paja CA, Spagnuolo G (2015) A genetic algorithm for identifying the single diode model parameters of a photovoltaic panel. Math Comput Simul 131:38–54. https://doi.org/10.1016/j.matcom.2015.10.008

15. Khare A, Rangnekar S (2013) A review of particle swarm optimization and its applications in solar photovoltaic system. Appl Soft Comput

16. Bana S, Saini RP (2017) Identification of unknown parameters of a single diode photovoltaic model using particle swarm optimization with binary constraints. Renew Energy 101:1299–1310. https://doi.org/10.1016/j.renene.2016.10.010

17. Oliva D, Cuevas E, Pajares G (2014) Parameter identification of solar cells using artificial bee colony optimization. Energy 72:93–102. https://doi.org/10.1016/j.energy.2014.05.011

18. Wang R, Zhan Y, Zhou H (2015) Application of artificial bee colony in model parameter identification of solar cells. Energies 8:7563–7581. https://doi.org/10.3390/en8087563

19. Jack V, Salam Z, Ishaque K (2016) An accurate modelling of the two-diode model of PV module using a hybrid solution based on differential evolution. Energy Convers Manag 124:42–50. https://doi.org/10.1016/j.enconman.2016.06.076

20. Abido MA, Khalid MS (2017) Seven-parameter PV model estimation using differential evolution. Electr Eng 1–11. https://doi.org/10.1007/s00202-017-0542-2

21. Askarzadeh A (2013) A discrete chaotic harmony search-based simulated annealing algorithm for optimum design of PV/wind hybrid system. Sol Energy

22. Askarzadeh A, Rezazadeh A (2012) Parameter identification for solar cell models using harmony search-based algorithms. Sol Energy

23. Jovanovic R, Kais S, Alharbi FH (2016) Cuckoo search inspired hybridization of the Nelder-Mead simplex algorithm applied to optimization of photovoltaic cells. Appl Math Inf Sci 10:961–973. https://doi.org/10.18576/amis/100314

24. Ma J, Ting TO, Man KL et al (2013) Parameter estimation of photovoltaic models via cuckoo search. J Appl Math 2013:1–8. https://doi.org/10.1155/2013/362619

25. Humada AM, Hojabri M, Mekhilef S, Hamada HM (2016) Solar cell parameters extraction based on single and double-diode models: a review. Renew Sustain Energy Rev 56:494–509. https://doi.org/10.1016/j.rser.2015.11.051

26. Tamrakar R, Gupta A (2015) A review: extraction of solar cell modelling parameters 3. https://doi.org/10.17148/IJIREEICE.2015.3111

27. Chan DSH, Phillips JR, Phang JCH (1986) A comparative study of extraction methods for solar cell model parameters. Scopus

28. Karaboga D (2005) An idea based on honey bee swarm for numerical optimization. Tech Rep TR06, Erciyes Univ 10. https://doi.org/citeulike-article-id:6592152

29. Storn R, Price K (1997) Differential evolution—a simple and efficient heuristic for global optimization over continuous spaces. J Glob Optim 341–359. https://doi.org/10.1023/A:1008202821328

30. Geem ZW (2001) A new heuristic optimization algorithm: harmony search. Simulation

31. Rashedi E, Nezamabadi-pour H, Saryazdi S (2009) GSA: a gravitational search algorithm. Inf Sci (Ny) 179:2232–2248. https://doi.org/10.1016/j.ins.2009.03.004

32. Yang XS, Deb S (2009) Cuckoo search via Lévy flights. 2009 World Congr Nat Biol Inspired Comput NABIC 2009—Proc 210–214. https://doi.org/10.1109/NABIC.2009.5393690

33. Civicioglu P (2012) Transforming geocentric Cartesian coordinates to geodetic coordinates by using differential search algorithm. Comput Geosci 46:229–247. https://doi.org/10.1016/j.cageo.2011.12.011

34. Askarzadeh A (2016) A novel metaheuristic method for solving constrained engineering optimization problems: crow search algorithm. Comput Struct 169:1–12. https://doi.org/10.1016/j.compstruc.2016.03.001
35. Hansen N, Ostermeier A, Adapting arbitrary normal mutation distributions in evolution strategies: the covariance matrix adaptation. In: Proceedings of IEEE international conference on evolutionary computation. IEEE, pp 312–317
36. Hansen N, Ostermeier A (2001) Completely derandomized self-adaptation in evolution strategies. Evol Comput 9:159–195
37. Kennedy J, Eberhart R (1995) Particle swarm optimization. Neural Netw, 1995 Proc, IEEE Int Conf 4:1942–1948. https://doi.org/10.1109/ICNN.1995.488968
38. Paenke I, Jin Y, Branke J (2009) Balancing population- and individual-level adaptation in changing environments. Adapt Behav 17:153–174. https://doi.org/10.1177/1059712309103566
39. Patwardhan AP, Patidar R, George NV (2014) On a cuckoo search optimization approach towards feedback system identification. Digit Signal Process A Rev J 32:156–163. https://doi.org/10.1016/j.dsp.2014.05.008
40. Barthelemy P, Bertolotti J, Wiersma DS (2008) A Lévy flight for light. Nature 453:495–498. https://doi.org/10.1038/nature06948
41. Yona A, Senjyu T, Funabshi T, Sekine H (2008) Application of neural network to 24-hours-ahead generating power forecasting for PV system. IEEJ Trans Power Energy 128:33–39. https://doi.org/10.1541/ieejpes.128.33
42. Hiyama T, Kouzuma S, Imakubo T (1995) Identification of optimal operating point of PV modules using neural network for real time maximum power tracking control. IEEE Trans Energy Convers 10:360–367. https://doi.org/10.1109/60.391904
43. Karatepe E, Boztepe M, Colak M (2006) Neural network based solar cell model. Energy Convers Manag 47:1159–1178. https://doi.org/10.1016/j.enconman.2005.07.007
44. Ishaque K, Salam Z, Taheri H (2011) Simple, fast and accurate two-diode model for photovoltaic modules. Sol Energy Mater Sol Cells 95:586–594. https://doi.org/10.1016/j.solmat.2010.09.023
45. Ji Y-H, Kim J-G, Park S-H et al, C-language based PV array simulation technique considering effects of partial shading
46. Beyer H-G (1999) Evolutionary algorithms in noisy environments: theoretical issues and guidelines for practice. https://doi.org/10.1016/S0045-7825(99)00386-2
47. Jun-hua L, Ming L (2013) An analysis on convergence and convergence rate estimate of elitist genetic algorithms in noisy environments. Opt—Int J Light Electron Opt 124:6780–6785. https://doi.org/10.1016/j.ijleo.2013.05.101
48. Ma T, Yang H, Lu L (2014) Solar photovoltaic system modeling and performance prediction. Renew Sustain Energy Rev 36:304–315. https://doi.org/10.1016/j.rser.2014.04.057
49. Nishioka K, Sakitani N, Uraoka Y, Fuyuki T (2007) Analysis of multicrystalline silicon solar cells by modified 3-diode equivalent circuit model taking leakage current through periphery into consideration. Sol Energy Mater Sol Cells 91:1222–1227. https://doi.org/10.1016/j.solmat.2007.04.009
50. Ishaque K, Salam Z (2011) An improved modeling method to determine the model parameters of photovoltaic (PV) modules using differential evolution (DE). Sol Energy 85:2349–2359. https://doi.org/10.1016/j.solener.2011.06.025
51. Abdul Hamid NF, Rahim NA, Selvaraj J (2013) Solar cell parameters extraction using particle swarm optimization algorithm. In: 2013 IEEE conference on clean energy and technology (CEAT). IEEE, pp 461–465
52. Hochberg Y (1988) A sharper Bonferroni procedure for multiple tests of significance. Biometrika 75:800–802. https://doi.org/10.1093/biomet/75.4.800
53. Armstrong RA (2014) When to use the Bonferroni correction. Ophthalmic Physiol Opt 34:502–508. https://doi.org/10.1111/opo.12131
54. Cuevas E, González A, Fausto F, Zaldívar D, Pérez-Cisneros M (2015) Multithreshold segmentation by using an algorithm based on the behavior of Locust Swarms. Math Probl Eng 2015:805357

Chapter 8
Comparison of Metaheuristics Techniques and Agent-Based Approaches

Agent-based models represent new approaches to characterize systems through simple rules. Under such techniques, complex global behavioral patterns emerge from the agent interactions produced by the rules. Recently, due to their capacities, metaheuristic methods have attracted the attention of the optimization community. Even though these approaches are built emulating very distinct phenomena, their structure and operators are very similar. Despite their diversity, several metaheuristic methods recycle the same elements from other metaheuristic techniques that has demonstrated to be efficient. These elements have been designed without considering the produced final search patterns. Contrarily, agent-based modeling aims to relate the global behavioral patterns produced by the collective interaction of the individuals with the set of rules that describe their behavior. In this chapter, we remark the association between metaheuristic elements and agent-based models. Therefore, different agent-based structures that produce interesting search patterns can be employed to generate promising optimization methods. To demonstrate the abilities of this methodology, an agent-based approach known as "Heroes and Cowards" has been structured as a metaheuristic technique. This agent-based model implements a small set of rules to generate two search patterns that can perform as exploration and exploitation stages. To evaluate its performance, this algorithm has been compared with several metaheuristic methods.

8.1 Introduction

Since real-world processes become more interconnected, simple models are no longer enough to analyze them. The wide availability of fast computing resources has allowed the construction and analysis of more complex models. Under such conditions, it has emerged a new field of knowledge known as complex systems [1]. In complex systems, it is studied how systems affect individual behaviors, especially when such individuals have the capacity to influence these systems. In these

© The Author(s), under exclusive license to Springer Nature Switzerland AG 2023
E. Cuevas et al., *Analysis and Comparison of Metaheuristics*, Studies in Computational
Intelligence 1063, https://doi.org/10.1007/978-3-031-20105-9_8

systems, complex behaviors of higher-level organizations appear as a consequence of the collective interaction of individuals that participate in a self-organizing process [2].

Agent-based modeling [3] represents a new paradigm in artificial intelligence to model complex systems using agents or elements. Agents maintain behaviors that are described by simple rules and influenced by the collective interaction with other agents. Under this paradigm, global behavioral patterns that have not been directly programmed emerge through the collective interaction among agents. Agent-based models attempt to relate how global regularities may emerge through processes of collective cooperation. Under this scheme, a population of agents maintains a behavior characterized by a set of simple rules. The objective of such rules is to emulate the individual movements of real actors when they interact with their local environment. Although the system is modeled from the individual point of view, its main properties are visualized from a global perspective. The powerful modeling characteristics of the Agent-based models have motivated their use in several applications which include the prediction of the spread in epidemics [4], the behavior in supply chains [5] and the stock market [6], the characterization of the immune system [7], the understanding about the fall of ancient civilizations [8], the consumer purchasing behavior [9], to name a few.

Under the agent-based methodology, several interesting basic global patterns have been proposed to simulate complex phenomena such as diffusion, concentration and insolating, fire spreading, segregation and others. These behavioral patterns have been analyzed in terms of the simple rules that provoke them. In the complex system community, there is a model known as "Heroes and Cowards" [1, 10–12] used to illustrate how simple rules can produce complex collective behaviors that are very difficult to reproduce by employing classical modeling techniques. The model produces complex global patterns of concentration and distribution through the interaction of agents that follows simple behavioral rules. In Heroes and Cowards, each agent selects other agent as its "friend" and another as its "enemy". The model consists of two phases. In the first stage, every agent behaves as "coward". Under this condition, the agent moves so that the friend is always located between it and its enemy (in a cowardly manner, hiding from its enemy behind its friend). During this phase, agents are distributed within the space as a consequence of the scape process from the enemy. In the second phase, each agent behaves as "hero". Therefore, the agent presents a strategy where it is moved in a position between its friend and enemy (protecting the friend from the enemy in a heroically). During this stage, agents concentrate around positions marked by the agent distributions.

On the other hand, metaheuristic schemes are abstract optimization models that emulate several biological or social systems. Under these methods, individuals are initially generated by using random values where each individual represents a possible solution actually. An objective function evaluates the quality of each individual in terms of its solution. With this information, at each step, individuals are modified according to some rules that define their behavior within the search space. Such rules correspond to abstract models supposedly extracted from natural or social mechanisms [13]. This process is repeated until a stop criterium has been reached. Metaheuristic methods have demonstrated their superiority in several real-world applications in situations where classical techniques cannot be used. In general terms, there is not a strict division of metaheuristic schemes. However, different categories of methods have been popularly accepted depending on some criteria such as cooperation among the agents, the source of inspiration, or the type of operators [14, 15]. Metaheuristic schemes are divided regarding their inspiration into three classes: swarm-based, physics-based and evolution-based. Swarm-inspired metaheuristic methods employ behavioral models obtained from the cooperative interaction of several species of insects or animals to design a search strategy. Currently, a high number of swarm-based optimization methods have been introduced in the literature. Some of the most popular swarm schemes involve, Best-so-far Artificial Bee Colony (BSF ABC) [16], Particle Swarm Optimization (PSO) algorithm [17–19], Social Spider Optimization (SSO) [20], Crow Search Algorithm (CSA) [21, 22], Gray Wolf Optimizer (GWO) [23], Bat Algorithm (BA) [24], Cuckoo Search (CS) [25], Firefly Algorithm (FA) [26–28], to name a few. Physics-based metaheuristic approaches consider simplified physical formulations as models to modify individual positions from an iteration to another. Some of the most representative physics-based metaheuristic schemes include Gravitational Search Algorithm (GSA) [29], Simulated Annealing (SA) algorithm [30–32], Electromagnetism-like Mechanism (EM) [33], Big Bang-Big Crunch (BB-BC) [34], States of Matter Search (SMS) [35, 36] and Water Cycle Algorithm (WCA) [37]. Evolution-based schemes represent the oldest and most consolidate metaheuristic approaches that use evolution models as operators to influence the interaction among individuals. Therefore, elements such as mutation, reproduction, selection and recombination are employed to produce behavioral rules in this kind of method. Some examples of evolution-based metaheuristic methods include the Genetic Algorithm (GA) [38], Evolutionary Strategies (ES) [39–41], Self-Adaptive Differential Evolution (JADE) [42] and Differential Evolution (DE) [43].

Although all metaheuristic schemes emulate very different processes or systems, the operators used to model individual behavior are very similar. The idea behind the design of many metaheuristic methods is to configure a recycled set of rules that have demonstrated to be successful in previous approaches for producing new optimization schemes. Such common rules have been designed without considering the final global pattern obtained by the individual interactions. Different from metaheuristics, agent-based modeling aims to relate the global behavioral patterns produced by the collective interaction of the individuals with the set of rules that describe their behavior. Under this perspective, several agent-based modeling techniques that generate very complex search global behaviors can be used to produce or improve efficient optimization algorithms. In this chapter, we highlight the relationship between metaheuristic schemes and agent-based modeling. This chapter has two objectives: (I) To demonstrate the efficacy of agent-based models as metaheuristic methods; and (II) to show the promising potential in the combination of both artificial intelligence paradigms. In order to show the capacities of this association, the agent-based model of "Heroes and Cowards" is implemented as a metaheuristic method. To evaluate its performance, the presented algorithm has been tested in a set of 23 benchmark functions, including multimodal, unimodal, and hybrid benchmark functions. The competitive results indicate that even though agent-based modeling and metaheuristic schemes refer to distinct scientific communities, metaheuristic methods can increase their capabilities through the incorporation of concepts, formalisms, and models extracted from agent-based techniques.

This chapter is organized as follows: In Sect. 8.2, the basic concepts of agent-based modeling are introduced. In Sect. 8.3, the model of Heroes and Cowards is reviewed. In Sect. 8.4, the presented metaheuristic method is exposed. In Sect. 8.5 the experimental results and the comparative analysis is presented. Finally, in Sect. 8.6, conclusions are drawn.

8.2   Agent-Based Approaches

Agent-based modeling corresponds to a new scheme for simulating systems with interacting autonomous elements. Agents are artificial individuals programmed to perform pre-defined operations [44]. While they operate based on their own behavior, collaborate or compete with each other agents. The complexity of the actions conducted by an agent is quite simple. They range from elementary decisions (yes or no) to movements.

Agents interact in an environment (virtual map) in the form of a lattice or a multi-dimensional space. Agents can move freely within the environment. With this characteristic, it is possible to visualize the agent behaviors as a physical system, such as simulations of evacuations, traffic, biological systems, etc.

Most of the agent-based models are quite simple. They do not use sophisticated architectures or difficult behavioral rules. In spite of these simple models, they are capable of generating several global patterns (behaviors) as a consequence of the modeling characteristics produced by the interactions of a set of simple agents. Global behavioral patterns refer to consistent macroscopic regularities, such as coherent temporal, spatial and behavioral structures, or identifiable distributions.

A general agent-based modeling scheme consists of the following steps. First, a set of A agents $\{\mathbf{a}_1, \ldots, \mathbf{a}_A\}$ are initialized. Under this stage, agents are configured in a determined position or in a specific state. Then, each agent $\mathbf{a}_i$ ($i \in 1, \ldots, A$) is selected randomly or considering a particular order. For this agent $\mathbf{a}_i$, a set of rules are applied in order to change its position, state or relationship with other agents. These rules consider a relation of conditions imposed by other agents (specific agents) or local influences (neighbor agents). This process is repeated until a determined stop criterion has been reached.

Under the agent-based methodology, several interesting basic global patterns have been proposed to simulate complex phenomena such as diffusion, concentration and insolating, fire spreading, segregation and others. These behavioral patterns have been analyzed in terms of the simple rules that provoke them. In order to illustrate this methodology, two simple examples are considered: Fire spreading and segregation.

8.2.1 Fire Spreading

In fire spreading [1], the objective is to emulate the way in which fire moves through a zone of trees distributed with different densities. Fire, in the real world, does not spread under deterministic principles. From one tree to another tree, fire is transmitted based on a variety of elements such as the type of wood, wind, and how close the branches are to each other. Agent-based modeling allows emulating systems with many interconnected factors that affect a process. Under agent-based modeling, the fire spreading phenomenon is simulated by considering the following procedure. First, a lattice of $M \times N$ agents $\{a_1, \ldots, a_{M \times N}\}$ is randomly initialized. Each agent represents a tree or an empty space. One random agent a_R of the lattice is considered the location in which starts the fire. Considering this agent a_R, the next rule is applied. In the neighborhood of a_R, a new agent a_S is randomly selected. Then, with a probability p it is assumed that the fire is transmitted to a_S. Otherwise, the tree remains unburned. This process is repeated until all possible trees become burned.

8.2.2 Segregation

Schelling proposed an agent-based model to emulate the segregation phenomenon [45, 46] with the objective of providing an explanation for why people that maintain different ethnic origins tend to segregate geographically. In the model, two different types of agents $\mathbf{A} = \{a_1, \ldots, a_{A/2}\}$ and $\mathbf{B} = \{b_1, \ldots, b_{A/2}\}$ are randomly distributed in a finite two-dimensional space. In each step, the following rule is considered. A random agent from $\mathbf{A}$ or $\mathbf{B}$ is selected. Then, it is counted the number of agents of the same type around its neighborhood. If the fraction of agents of the same type is below a threshold Th, it moves to another position randomly chosen in the space.

The rule for this model is quite simple. Schelling discovered with this model how high the Th value had to be in order to occur segregation. It seems reasonable to assume that segregation requires high homophilic characteristics (high Th values). However, Schelling demonstrated that segregation phenomenon could happen with much lower Th values.

8.3 Heroes and Cowards Concept

In the complex system community, there is a model known as "Heroes and Cowards" [1, 10–12] used to illustrate how simple rules can produce complex collective behaviors that are very difficult to reproduce by employing classical modeling techniques. The model produces complex global patterns of concentration and distribution through the interaction of agents that follows simple behavioral rules.

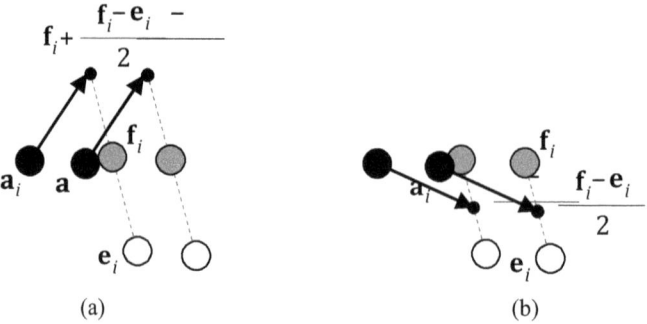

Fig. 8.1 Effect of the rules that control **a** the coward and **b** heroical behavior

In Heroes and Cowards, a set of A agents $\{\mathbf{a}_1, \ldots, \mathbf{a}_A\}$ are initialized with a random position in a two-dimensional space. Each agent $\mathbf{a}_i$ selects other agent $\mathbf{a}_p$ as its "friend" $\mathbf{f}_i$ and another $\mathbf{a}_q$ as its "enemy" $\mathbf{e}_i$ where $i, p, q \in (1, \ldots, A)$ and $i \neq p \neq q$. The friend $\mathbf{f}_i$ and enemy $\mathbf{e}_i$ selected by $\mathbf{a}_i$ maintain this association during the complete simulation. The model consists of two phases. In the coward stage, every agent behaves as "coward". Under this condition, the agent $\mathbf{a}_i$ moves so that the friend $\mathbf{f}_i$ is always located between $\mathbf{a}_i$ and the enemy $\mathbf{e}_i$ (in a cowardly manner, hiding from its enemy behind its friend). The rule that controls this behavior is formulated as follows:

$$\mathbf{a}_i(t+1) = (1 - \beta) \cdot \mathbf{a}_i(t) + \beta\left(\mathbf{f}_i + \left(\frac{\mathbf{f}_i - \mathbf{e}_i}{2}\right)\right) \tag{8.1}$$

where t corresponds to the current iteration and β ($\beta \in [0, 1]$) refers to a factor that determines the velocity with which the agent $\mathbf{a}_i$ is displaced. This behavior is illustrated in Fig. 8.1a.

In the heroical phase, each agent behaves as "hero". Therefore, the agent presents a strategy where $\mathbf{a}_i$ is moved in a position between its friend $\mathbf{f}_i$ and enemy $\mathbf{e}_i$ (protecting the friend from the enemy in a heroically). The behavioral rule that determines this interaction is formulated as follows:

$$\mathbf{a}_i(t+1) = (1 - \beta) \cdot \mathbf{a}_i(t) + \beta\left(\frac{\mathbf{f}_i - \mathbf{e}_i}{2}\right) \tag{8.2}$$

This behavior illustrated in Fig. 8.1b. The Heroes and Cowards model considers an artificial moderator that decides according to a specific number of iterations N_t when the agents behave as cowards (Eq. (8.1)) or heroes (Eq. (8.2)). Therefore, in the model, the phases are performed intercalated. N_t iterations last the coward phase while the heroical phase considers the next N_t iterations. This process continues until a determined number of phases have been reached.

Like any other agent-based model, the agents in Heroes and Cowards updates their position (state) in each iteration. Under such conditions, the relation of each agent $\mathbf{a}_i$ with its related friend $\mathbf{f}_i$ and enemy $\mathbf{e}_i$ is dynamic. Therefore, the model produces complex spatial behaviors of concentration and distribution through the interaction of all agents. During the coward phase, agents are distributed along the space as a consequence of the scape process from the enemy. On the other hand, during the hero stage, agents semi-concentrate around positions marked by the agent distributions.

Figure 8.2 presents examples of global patterns produced by the operation of the Heroes and Cowards model. In the Figure, the model is simulated by using a set of 50 agents ($A = 50$). Figure 8.2b shows the final obtained pattern by the model in its coward phase after 100 iterations considering as initial configuration the distribution shown in Fig. 8.2a. As it can be seen, agents, initially concentrated in the center, are distributed themselves along the space. Figure 8.2d represents the final obtained behavioral pattern by the heroical phase after 100 iterations considering as initial configuration the distribution shown in Fig. 8.2c. A simple inspection from the Figure indicates that agents originally distributed in the space make semi concentrations in regions of the two-dimensional space.

8.4 An Agent-Based Approach as a Metaheuristic Method

Although all metaheuristic schemes emulate very different processes or systems, the operators used to model individual behavior are very similar. The idea behind the design of many metaheuristic methods is to configure a recycled set of rules that have demonstrated to be successful in previous approaches for producing new optimization schemes. Such common rules have been designed without considering the final global pattern obtained by the individual interactions.

Different from metaheuristics, agent-based modeling aims to relate the global behavioral patterns produced by the collective interaction of the individuals with the set of rules that describe their behavior. Under this perspective, several agent-based modeling techniques that generate very complex search global behaviors can be used to produce or improve efficient optimization algorithms.

In this chapter, we highlight the relationship between metaheuristic schemes and agent-based modeling. In order to show the capacities of this association, the agent-based model of "Heroes and Cowards" is implemented as a metaheuristic method. The section is divided into three parts: (Sect. 8.4.1) Problem formulation, (Sect. 8.4.2) the description of the agent-based model of heroes and cowards as a metaheuristic method and (Sect. 8.4.3) the computational procedure.

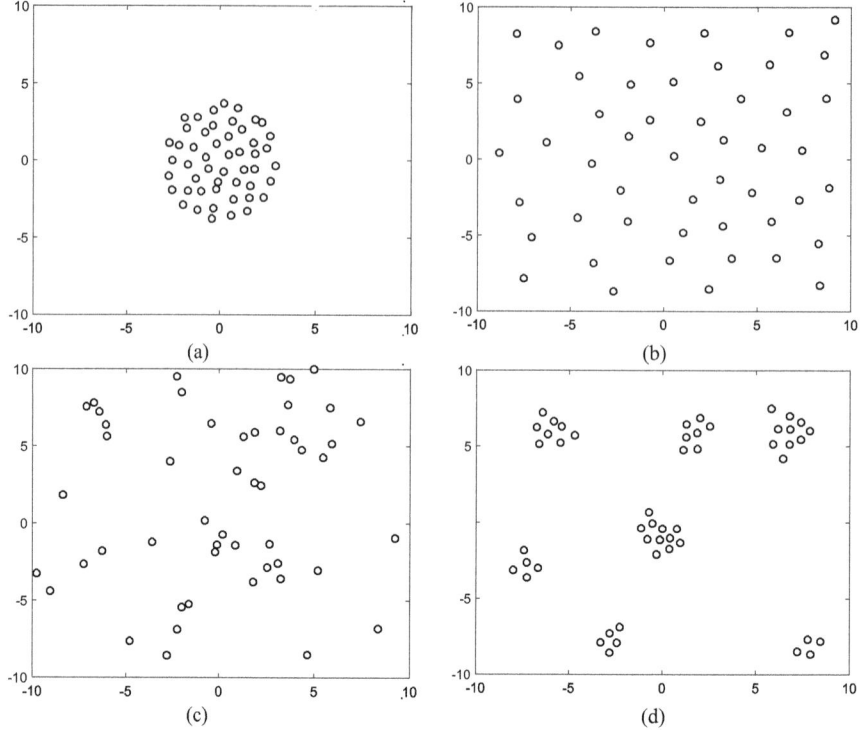

Fig. 8.2 Examples of global patterns produced by the operation of the Heroes and Cowards model. **a** Initial distribution, **b** global pattern produced by the model in the coward phase considering as initial configuration the provided by **a**, **c** initial distribution and **d** final pattern generated by the model in the heroical phase assuming as an initial distribution the agents shown in **c**

8.4.1 Problem Formulation

An optimization method is designed to find a global solution for a nonlinear problem with box constraints according to the following formulation [47].

$$\text{Maximize/minimize} \quad J(\mathbf{x})\mathbf{x} = (x_1, \dots, x_d) \in \mathbb{R}^d$$
$$\text{subject to} \qquad\qquad \mathbf{x} \in \mathbf{X} \tag{8.3}$$

where $J : \mathbb{R}^d \rightarrow \mathbb{R}$ corresponds to a d-dimensional nonlinear function and $\mathbf{X}$ represents a constrained search space $\left(\mathbf{x} \in \mathbb{R}^d | l_i \leq x_i \leq u_i, i = 1, \dots, d\right)$ by the lower (l_i) and upper (u_i) bounds. To solve the optimization problem formulated by Eq. 8.3, from a metaheuristic perspective, a population of $\mathbf{A}^k$ $\left(\{\mathbf{a}_1^k, \dots, \mathbf{a}_N^k\}\right)$ of N candidate solutions (agents) evolves from a starting point ($k=1$) to a **gen** number of iterations ($k = gen$). In the population, each agent $\mathbf{a}_i^k$ ($i \in [1, \dots, N]$) represents a d-dimensional vector $\{a_{i,1}^k, \dots, a_{i,d}^k\}$, which corresponds to the decision variables involved by the optimization problem. In the first iteration, the metaheuristic method starts generating a group of N agents with values uniformly distributed within the pre-specified lower (l_i)) and upper (u_i) bounds. Then, at each iteration, a determined number of metaheuristic operations are applied over the agents of the population $\mathbf{A}^k$ to produce the new population $\mathbf{A}^{k+1}$. The quality of each individual $\mathbf{a}_i^k$ is evaluated in terms of its solution regarding the objective function $J(\mathbf{a}_i^k)$ whose result represents the fitness value of $\mathbf{a}_i^k$. As the search strategy evolves, the best current agent $\mathbf{b}$ $\{b_1, \ldots, b_d\}$ is preserved, since $\mathbf{b}$ corresponds to the best available solution seen so-far.

8.4.2 Heroes and Cowards as a Metaheuristic Method

This subsection explains the way in which the agent-based approach of "Heroes and cowards" has been adapted to perform as a competitive optimization method. The method considers three elements: (Sect. 8.4.2.1) Initialization, (Sect. 8.4.2.2) operators and (Sect. 8.4.2.3) phase management.

8.4.2.1 Initialization

In the first iteration ($k = 1$), the method starts generating a set of A agents $\mathbf{A}^k = \{\mathbf{a}_1^k, \ldots, \mathbf{a}_N^k\}$ with random positions in a d-dimensional space $(\mathbf{a}_i^k = \{a_{i,1}^k, \ldots, a_{i,d}^k\})$. In this process, each decision variable $a_{i,j}^k$ ($i \in 1, \ldots, N$; $j = 1, \ldots, d$) that corresponds to the j-th parameter of the i-th agent is set with a numerical value uniformly determined between the defined lower (l_i) and upper (u_i) limits, so that

$$a_{i,j}^k = l_i + \mathrm{rand}(0, 1) \cdot (u_i - l_i) \tag{8.4}$$

Each agent $\mathbf{a}_i^k$ selects other agent $\mathbf{a}_p^k$ as its "friend" $\mathbf{f}_i^k$ and another $\mathbf{a}_q^k$ as its "enemy" $\mathbf{e}_i^k$ where $i, p, q \in (1, \ldots, A)$ and $i \neq p \neq q$. The friend $\mathbf{f}_i^k$ and enemy $\mathbf{e}_i^k$ selected by $\mathbf{a}_i^k$ maintain this association during the complete process.

8.4.2.2 Operators

The "Heroes and cowards" model consists of two processes: Coward and heroical phases. Under such conditions, the agent-based metaheuristic scheme implements an operator for each phase. Each operator updates the position of an agent $\mathbf{a}_i^k$ in relation to the position of its friend $\mathbf{f}_i^k = \{f_{i,1}^k, \ldots, f_{i,d}^k\}$ and enemy $\mathbf{e}_i^k = \{e_{i,1}^k, \ldots, e_{i,d}^k\}$. Such operations are practically the same as the behavioral rules involved in the original model, with only an adaptation. This modification represents the incorporation of a random number in order to add a stochastic effect in the search strategy. Therefore, the coward operator is defined as follows:

$$a_{i,j}^{k+1} = (1 - \beta) \cdot a_{i,j}^k + \beta \left(f_{i,j}^k + \left(\frac{f_{i,j}^k - e_{i,j}^k}{2} \right) \right) + \alpha \cdot \mathrm{rand}(-1, 1) \qquad (8.5)$$

where α represents the intensity of the stochastic effect and $\mathrm{rand}(-1, 1)$ a function that delivers a random number uniformly distributed between -1 and 1. On the other hand, the heroical operator is modeled as follows:

$$a_{i,j}^{k+1} = (1 - \beta) \cdot a_{i,j}^k + \beta \cdot \left(\frac{f_{i,j}^k - e_{i,j}^k}{2} \right) + \alpha \cdot \mathrm{rand}(-1, 1) \qquad (8.6)$$

In the adaptation of the heroes and coward model as an optimization scheme, the location of all agents is updated by the coward or heroical operators except for the best element $\mathbf{b}$ of the population $\mathbf{A}^k$. In case of a maximization problem, this agent will be selected in each iteration k so that

$$\mathbf{b} = \arg\max_{\mathbf{a}_i^k \in \mathbf{A}^k} J(\mathbf{a}_i^k) \qquad (8.7)$$

Once obtained, this agent $\mathbf{b}$ is not modified by the operators.

8.4.2.3 Phase Management

The position of each agent $\mathbf{a}_i^k$ (except $\mathbf{b}$) is modified iteratively according to one of the operators while the phase of the model has not been changed. Similar to the original model, the phases are performed intercalated. N_t iterations last the coward phase (Eq. (8.5)) while the heroical phase (Eq. (8.6)) considers the next N_t iterations. This process continues until a stop criterion has been reached.

8.4.3 Computational Procedure

The adapted "Heroes and cowards" model has been implemented as an iterative scheme that considers some processes in its operation. In the form of pseudo-code, Algorithm 8.1 summarizes the operations of the whole process. The approach requires as input data the number of agents A, the displacement velocity β, the intensity of the stochastic effect α, the number of iterations of each phase N_t and the maximum number of executions gen (line 1). As another metaheuristic scheme, initially (line 2), the method produces a set of (A) agents with positions uniformly distributed between the pre-specified limits. Such agents correspond to the initial population $\mathbf{A}^1$. Then, the best element $\mathbf{b}$ from $\mathbf{A}^1$ regarding its fitness value is selected

(Line 3). The method begins considering N_t iterations of the coward phase (line 4). Therefore, during this phase each agent a_i^k except the best element **b** is modified (line 7) by the operator defined in Eq. (8.5). Once the N_t iterations have been reached, (line 14) a change of phase is achieved (line 15). After changed the phase, the heroical phase is conducted. During this phase each agent a_i^k except the best element **b** is modified (line 10) by the operator defined in Eq. (8.6). After the application of an operator, the best element **b** from A^k regarding its fitness value is selected (Line 12). This process is conducted until the maximal number of generations gen has been reached. As output (line 19), the algorithm delivers as output the last obtained **b**, since it represents the final solution.

Algorithm 8.1 Summarized processes of the adapted "Heroes and cowards" model.

Algorithm 1. Pseudo-code for the "Heroes and cowards" model
1. **Input:** $A, \beta, \alpha, N_t, gen$
2. $A^1 \leftarrow$ **Initialize**(A);
3. **b** $\leftarrow$**SelectBestAgent**(A^1);
4. Phase= Coward; m=1;
5. **while** $k <= gen$ **do**
6. **If** (Phase== Coward)
7. $A^{k+1} \leftarrow$ **CowardOperator**($\forall a_i^k \in A^k$where $a_i^k \neq$ **b**);
8. **end if**
9. **If** (Phase== Heroical)
10. $A^{k+1} \leftarrow$ **HeroicalOperator**($\forall a_i^k \in A^k$where $a_i^k \neq$ **b**);
11. **end if**
12. **b** $\leftarrow$**SelectBestAgent**(A^{k+1});
13. m=m+1; $k = k + 1$;
14. **If** (m== N_t)
15. Phase$\leftarrow$ **ChangeOfPhase**;
16. m=1;
17. **end if**
18. end **while**
19. **Output: b**

In order to illustrate the operation of the "Heroes and cowards" model, an optimization example is carried out. The example aims to detect the maximal value of the two-dimensional objective function $J(x_1, x_2)$ defined in Eq. (8.8).

$$J(x_1, x_2) = 3(1 - x_1)^2 e^{-(x_1^2 - x_2^2)} - 10\left(\frac{x_1}{5} - x_1^3 - x_2^5\right)e^{(-x_1^2 - x_2^2)} - 1/3e^{(-(x_1+1)^2 - x_2^2)}$$

$$(8.8)$$

In the example, the algorithm has been set with the following parameters: $A = 20$, $\beta = 0.7, \alpha = 0.3, N_t = 20, gen = 100$. Assuming these parameters, all the agents

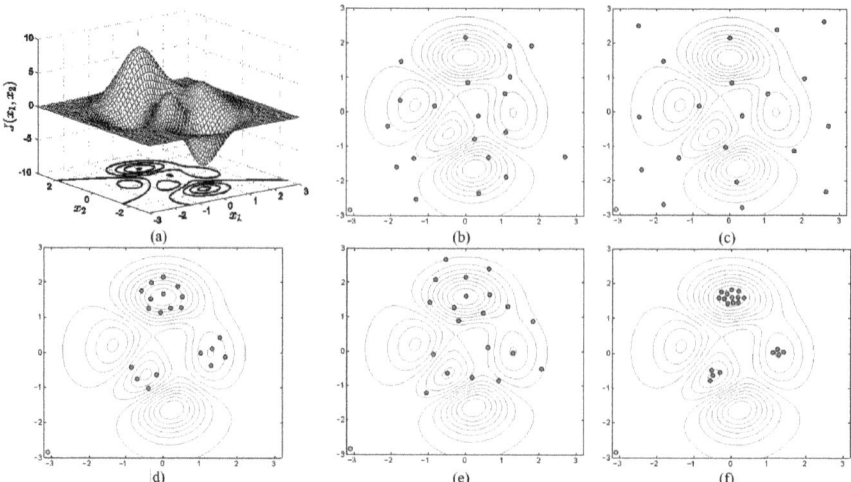

Fig. 8.3 "Heroes and cowards" model behavior considering 100 iterations when it solves the optimization problem formulated by Eq. (8.8). **a** Objective function with its respective contour, **b** 1 iteration (initialization), **c** 20 iterations (first coward phase), **d** 40 iterations (First heroical phase), **e** 60 iterations (second coward phase) and **f** 80 iterations (second heroical phase)

$\{\mathbf{a}_i^1 = (a_{i,1}^1, a_{i,2}^1)\}$ from $\mathbf{A}^1$ are initialized with random values uniformly distributed within the interval of $3 \leq x_1, x_2 \leq 3$. In the coward phase of the search strategy, the scheme promotes the distribution and exploration of solutions along the search space. On the other hand, in the heroical phase, the semi concentration of solutions is produced through the use of attraction movements. As the iterations progress, in the heroical phase, the exploitation is intensified to refine the quality of their solutions.

Figure 8.3 shows the behavior of the "Heroes and cowards" model. Figure 8.3a exhibits the objective function to optimize $J(x_1, x_2)$ with its respective contour representation. During the evolution of the algorithm, five points, (b) 1 iteration (initialization), (c) 20 iterations (first coward phase), (d) 40 iterations (First heroical phase), (e) 60 iterations (second coward phase) and (f) 80 iterations (second heroical phase), have been selected to show its operation. Point (b) represents an early stage of the algorithm where the elements are almost in their initial random location. As can be seen in Fig. 8.3c–e, the model produces groups as the iterations increase, until all elements converge in Fig. 8.3f to the global and local maxima.

8.5 Comparison with Metaheuristic Methods

Metaheuristic optimization techniques have been proposed as stochastic algorithms to solve optimization problems where classical methodologies are not suitable to operate since real-world optimization problems contain multiple optima. To evaluate the performance of metaheuristic schemes, the scientific community has proposed a set of benchmark functions to numerically expose the performance of such methods. In this study, the performance of metaheuristic methodologies is evaluated over a standard set of 23 mathematical benchmark functions [47, 48]. Such a benchmark set contains functions with different complexities to measure the precision, robustness and scalability of those mechanisms. Additionally, three engineering design problems commonly found in the related literature [21, 49] are used to evaluate the capabilities of the agent-based method by solving real-world optimization scenarios. During the optimization process, the algorithms are evaluated considering the maximum number of generations (*gen*) as stop criterion. This criterion has been extensively used in the metaheuristic optimization domain.

This section presents the numerical results of the agent-based "Heroes and cowards" model, which for identification proposes will be called as EA-HC. This model, as a metaheuristic algorithm, is evaluated using a set of benchmark functions as well as engineering design optimization problems. In Appendix 8.1, Table 8.12 mathematically describes the set of test functions used in the performance analysis. In the table, n corresponds to the n-dimensional vector at which the test functions are evaluated, $f(x^*)$ represents the optimal value of a given function evaluated at position x^* and S corresponds to the search space, defined by the lower and upper limits of the search space. To prove the scalability of the agent-based method, the evaluation for each test function is operated by 30 and 100-dimensional search spaces.

The performance results exposed by the proposed method are compared against the performance results of 8 evolutionary methodologies, named; Artificial Bee Colony (ABC) [50], Differential Evolution (DE) [43], Particle Swarm Optimization (PSO) [17], Cuckoo Search (CS) [25], Differential Search (DS) [51], Moth-Flame Optimization (MFO) [52], Multi-Verse Optimizer (MVO) [53], and the Sine Cosine Algorithm (SCA) [54]. Also, the numerical comparison among EA-HC method and the rest of competitors considering design optimization problems is analyzed. Each real-world design problem includes the well-known Three-bar truss design, Tension/compression spring design and Welded beam design. In Appendix 8.2, the Tables 8.13, 8.14 and 8.15 mathematically describe each optimization problem, respectively.

The experimental results are divided into four sub-sections. In the first Sect. 8.5.1, the performance of the agent-based algorithm is evaluated with regard to its tuning parameters. In the second Sect. 8.5.2, the overall performance of the presented method is compared to different popular metaheuristic algorithms is provided.

In the third Sect. 8.5.3, the convergence analysis for each test function considering each metaheuristic approach is presented. Finally, in the fourth Sect. 8.5.4 the ability of the "heroes and coward" model to solve engineering problems is analyzed.

8.5.1   Performance Evaluation with Regard to Its Own Tuning Parameters

The two parameters β and α present a determinant influence in the expected performance of the EA-HC scheme. In this sub-section is analyzed the behavior of the agent-based scheme considering different configurations of these parameters. In the test, one factor-at-a-time of the two parameters is tested while the other element remains fixed to a default value. To minimize the stochastic effect, each benchmark function is executed independently for 30 times. As a termination criterion, the maximum number of iterations is considered. It has been set to 1000. In all simulations, the population size is fixed to 50 individuals.

First, the behavior of the proposed algorithm is analyzed, considering different values for β. In the analysis, the values of β are varied from 0 to 1, whereas the value of α remains fixed to 0.3. In the simulation, the proposed method is executed independently 30 times for each value of β, on each benchmark function. Then, the performance of the proposed algorithm is evaluated, considering different values for α. In the experiment, the values of α are varied from 0.0 to 0.5, whereas the value of β remains fixed to 0.7. The obtained results suggest that a proper combination of the parameter values can improve the performance of the proposed method and the quality of solutions. With the experiment can be concluded that the best parameter set is composed of the following values: $\beta = 0.7$ and $\alpha = 0.3$. They are kept for the next experiments.

8.5.2   Performance Comparison

In this section, the performance of EA-HC is analyzed and numerically compared in terms of the fitness value (in this study, it is considered the performance results for minimization) against nine well-known evolutionary approaches considering a set of 23 test functions. The selected test functions include uni-modal, multi-modal and hybrid benchmark functions. To make a fair comparison among evolutionary methods, the evolutionary process of each algorithm uses $gen = 1000$ as a stop condition. This condition has been chosen to maintain compatibility with the numerical results of most of the related and novel works [55–57] in the literature. To prove the scalability of the proposed methods, the simulations are evaluated in $n = 30$ and $n = 100$ dimensions and each experiment has been executed by 30 runs. Since metaheuristic methods are stochastic search methods, statistically validation for the results must also be included in order to eliminate the random effect. In this study, the numerical results have been validated, considering the Wilcoxon rank sum [58].

The performance of most of the metaheuristic approaches is given by the correct setting step of configuration parameters to improve their search capabilities. Such configurations inherently depend on the optimization problem that wanted to be solved. In the optimization theory, The No-Free-Lunch (NFL) theorem states that

Table 8.1 Parameter configuration for each metaheuristic method used in the experimental study

Method	Parameter(s)	References
ABC	The population size has been set as $limit = 50$	[50]
DE	The variant considered is DE/rand/bin, where crossover probability $cr =$ 0.5 and differential weight $dw = 0.2$	[43]
PSO	The parameters are set to $c_1 = 2$ and $c_2 = 2$ with linearly decreasing weight factor from 0.9 to 0.2	[25]
CS	The discover probability is set to $pa = 0.25$	[17]
DS	The algorithm has been implemented considering the scale parameter cp $= 0.5$ for the Gamma distribution	[51]
MFO	The source code has been obtained by its reference	[52]
MVO	The source code has been obtained by its reference	[53]
SCA	The implementation considers the guidelines described by the author	[54]
EA-HC	$\beta = 0.7, \alpha = 0.3$ and $N_t = 50$	

there is no single algorithm that can solve any optimization problem. That is, if an algorithm X outperforms algorithm Y for the W optimization problem, maybe algorithm Y outperforms algorithm X for the G optimization problem. Under such circumstances, the design of metaheuristic algorithms can include tuning parameters to increase the possibility of locating optima values efficiently.

In this comparison, the implementation of the parameter configuration for each algorithm has been set following the reported guidelines by the authors. Table 8.1 summarizes the configuration step for each metaheuristic method.

For the experimental study, the population size has been set to 50 individuals for each metaheuristic approach. Tables 8.2 and 8.3 contain the performance results of the numerical comparison. Table 8.2 reports the numerical results considering $n =$ 30 dimensions. Table 8.3 reports the numerical results for $n = 100$ dimensions. In the tables, the best fitness value is represented as f_{Best}, the average fitness as $\overline{f}$, the standard deviation as σ_f and the worst value as f_{Worst}. The tables also present the best performance entries in boldface.

According to Table 8.2, it is quite evident that the proposed EA-HC outperforms the rest of the metaheuristic algorithms considered in the comparison study. In case of functions f_6, f_8, f_9, f_{13}, f_{16}, f_{20}, and f_{22}, the proposed EA-HC performs quite similarly. For function f_6, ABC, CS, DS, MFO and SCA achieve similar performance than EA-HC. For function f_8, ABC, CS, DS, MVO and EA-HC obtain the same best fitness value however, it can be shown the median value of the fitness value for EA-HC is quite greater than the rest of median values in this function. Also, this phenomenon is replied in functions f_9 and f_{16}, where the performance of ABC, CS, and DS presents the same fitness value and ABC, DS, MFO and SCA achieve the same best fitness value then EA-HC, respectively. One of the most distinctive characteristics of EA-HC, is the process of changing between exploration and exploitation stages during the entire optimization process. As a consequence, the mean value of fitness is affected due to this changing mechanism. For the remaining functions, only ABC

Table 8.2 Minimization results of benchmark functions of Table 8.12 with $n = 30$

		ABC	DE	PSO	CS	DS	MFO	MVO	SCA	EA-HC
f_1	f_{Best}	1.05E−12	7.28E+03	8.23E+02	4.78E−03	1.88E−06	1.94E−06	1.13E−01	9.45E−07	**1.47E−144**
	f_{Worst}	7.08E−11	1.74E+04	3.32E+03	1.67E−02	2.40E−04	1.00E+04	2.92E−01	9.22E−02	5.47E−05
	$\overline{f}$	1.52E−11	1.23E+04	1.66E+03	9.27E−03	1.99E−05	1.20E+03	1.84E−01	6.52E−03	1.89E−06
	σ_f	1.72E−11	2.38E+03	6.65E+02	2.85E−03	4.32E−05	3.28E+03	4.55E−02	1.83E−02	9.98E−06
f_2	f_{Best}	4.14E−07	3.99E+01	2.31E+01	6.05E−01	7.66E−05	5.58E−05	1.98E−01	2.43E−09	**3.63E−77**
	f_{Worst}	2.22E−06	6.60E+01	1.03E+02	5.54E+00	1.27E−03	8.00E+01	5.45E−01	6.03E−05	4.39E−02
	$\overline{f}$	1.13E−06	5.20E+01	4.16E+01	1.67E+00	3.45E−04	2.85E+01	3.16E−01	6.33E−06	1.74E−03
	σ_f	4.77E−07	6.63E+00	1.89E+01	1.22E+00	2.78E−04	1.97E+01	8.69E−02	1.29E−05	8.01E−03
f_3	f_{Best}	6.00E+03	4.51E+04	2.13E+03	2.59E+02	9.98E+02	2.31E+02	9.17E+00	4.36E+01	**5.88E−40**
	f_{Worst}	1.56E+04	7.93E+04	2.03E+04	6.77E+02	4.97E+03	4.00E+04	3.64E+01	1.10E+04	1.44E+03
	$\overline{f}$	1.14E+04	5.79E+04	7.96E+03	4.37E+02	2.91E+03	1.56E+04	1.86E+01	3.48E+03	8.31E+01
	σ_f	2.75E+03	9.38E+03	3.64E+03	9.09E+01	1.28E+03	1.05E+04	7.64E+00	3.42E+03	3.11E+02
f_4	f_{Best}	1.83E−01	5.95E+01	1.50E+01	2.21E+00	2.76E+00	3.25E+01	3.11E−01	1.20E+00	**4.58E−140**
	f_{Worst}	5.60E+01	7.65E+01	3.50E+01	4.79E+00	9.96E+00	8.01E+01	1.14E+00	2.50E+01	9.81E−28
	$\overline{f}$	4.46E−01	6.86E+01	2.23E+01	3.39E+00	5.67E+00	5.65E+01	6.28E−01	1.15E+01	3.27E−29
	σ_f	8.00E+00	3.73E+00	4.30E+00	6.72E−01	1.90E+00	9.87E+00	2.07E−01	6.61E+00	1.79E−28
f_5	f_{Best}	1.91E−01	2.75E+01	1.40E+02	2.31E+01	2.24E+01	9.12E+00	2.34E+01	2.70E+01	**5.25E−03**
	f_{Worst}	2.72E+01	2.89E+01	1.05E+03	2.55E+01	7.89E+01	9.01E+04	8.14E+01	2.87E+01	2.87E+01
		2.44E+01	2.88E+01	3.68E+02	2.47E+01	2.66E+01	1.57E+04	3.03E+01	2.79E+01	2.17E+01
	σ_f	1.78E+00	4.66E−02	1.87E+02	5.80E−01	9.92E+00	3.39E+04	1.34E+01	4.46E−01	1.19E+01
f_6	f_{Best}	**0.00E+00**	8.45E+03	7.65E+02	**0.00E+00**	**0.00E+00**	**0.00E+00**	1.00E+00	**0.00E+00**	**0.00E+00**

(continued)

Table 8.2 (continued)

		ABC	DE	PSO	CS	DS	MFO	MVO	SCA	EA-HC
	f_{Worst}	0.00E+00	1.70E+04	3.34E+03	0.00E+00	0.00E+00	1.00E+04	1.70E+01	2.00E+00	1.00E+01
	$\bar{f}$	0.00E+00	1.19E+04	1.93E+03	0.00E+00	0.00E+00	1.60E+03	6.07E+00	6.67E-02	3.33E-01
	σ_f	0.00E+00	2.24E+03	6.78E+02	0.00E+00	0.00E+00	3.70E+03	3.12E+00	3.65E-01	1.83E+00
f_7	f_{Best}	9.31E-02	3.56E+00	2.16E+00	1.59E-02	1.41E-02	3.01E-02	6.10E-03	3.22E-03	**5.05E-06**
	f_{Worst}	3.07E-01	1.57E+01	1.14E+00	6.45E-02	6.47E-02	1.89E+01	3.71E-02	1.12E-01	2.14E-02
	$\bar{f}$	2.21E-01	9.27E+00	5.38E+00	3.36E-02	3.77E-02	1.84E+00	1.38E-02	2.48E-02	1.04E-03
	σ_f	5.28E-02	2.57E+00	2.34E+00	1.10E-02	1.42E-02	3.91E+00	6.42E-03	2.21E-02	3.89E-03
f_8	f_{Best}	**2.07E+00**	5.18E+00	4.63E+00	**2.07E+00**	**2.07E+00**	2.08E+00	**2.07E+00**	2.23E+00	**2.07E+00**
	f_{Worst}	2.07E+00	9.68E+00	1.25E+01	2.07E+00	2.08E++00	2.13E+00	2.07E+00	2.67E+00	3.25E+00
	$\bar{f}$	2.07E+00	7.40E+00	8.09E+00	2.07E+00	2.07E+00	2.10E+00	2.07E+00	2.36E+00	2.12E+00
	σ_f	2.65E-08	1.01E+00	1.93E+00	4.47E-06	2.60E-03	1.26E-02	6.13E-05	9.78E-02	2.19E-01
f_9	f_{Best}	**7.13E-02**	3.42E+00	6.05E-01	**7.13E-02**	**7.13E-02**	8.32E-02	7.15E-02	2.16E-01	**7.13E-02**
	f_{Worst}	7.13E-02	6.50E+00	2.60E+00	7.13E-02	7.40E-02	1.33E-01	7.16E-02	6.32E-01	6.51E+00
	$\bar{f}$	7.13E-02	4.49E+00	1.52E+00	7.13E-02	7.20E-02	9.99E-02	7.15E-02	3.57E-01	3.83E-01
	σ_f	1.06E-10	7.30E-01	6.29E-01	6.66E-07	1.13E-03	1.27E-02	4.37E-05	1.12E-01	1.26E+00
f_{10}	f_{Best}	8.20E-09	2.45E+02	9.56E+01	6.49E+01	2.56E+00	6.87E+01	5.78E+01	5.95E-06	**0.00E+00**
	f_{Worst}	1.02E+00	3.06E+02	2.42E+02	1.08E+02	1.83E+01	2.42E+02	2.09E+02	8.81E+01	2.53E+02
	$\bar{f}$	1.67E-01	2.75E+02	1.74E+02	8.52E+01	8.37E+00	1.43E+02	1.10E+02	2.02E+01	5.41E+01
	σ_f	3.79E-01	1.47E+01	3.12E+01	1.02E+01	2.59E+00	4.01E+01	3.58E+01	2.75E+01	9.09E+01
f_{11}	f_{Best}	2.04E-06	1.59E+01	7.60E+00	2.20E+00	2.14E-04	5.12E-04	1.03E-01	4.16E-05	**8.88E-16**
	f_{Worst}	3.54E-05	1.77E+01	1.47E+01	6.92E+00	4.89E-03	2.00E+01	2.61E+00	2.02E+01	4.67E-01

(continued)

Table 8.2 (continued)

		ABC	DE	PSO	CS	DS	MFO	MVO	SCA	EA-HC
f_{12}	$\bar{f}$	1.55E−05	1.67E+01	1.13E+01	3.91E+00	1.48E−03	1.39E+01	9.64E−01	1.64E+01	1.56E−02
	σ_f	8.37E−06	4.70E−01	1.66E+00	1.30E+00	1.31E−03	8.28E+00	7.69E−01	7.50E+00	8.53E−02
	f_{Best}	5.15E−11	7.27E+01	8.75E+00	4.79E−02	6.01E−07	5.09E−06	2.38E−01	2.26E−06	**0.00E+00**
	f_{Worst}	2.80E−05	1.79E+02	2.77E+01	1.81E−01	3.44E−02	9.09E+01	6.10E−01	6.93E−01	3.15E−03
	$\bar{f}$	1.20E−06	1.12E+02	1.73E+01	1.02E−01	7.08E−03	1.45E+01	4.47E−01	1.75E−01	1.05E−04
	σ_f	5.13E−06	2.24E+01	4.68E+00	2.77E−02	1.00E−02	3.35E+01	1.10E−01	2.00E−01	5.75E−04
f_{13}	f_{Best}	**−3.0E+00**	5.39E−02	3.24E+00	2.95E+00	**3.00E+00**	**3.00E+00**	2.85E+00	**3.00E+00**	**−3.00E+00**
	f_{Worst}	−3.00E+00	1.44E+00	3.92E+00	−2.42E+00	−3.00E+00	−6.00E−01	−1.08E+00	−3.00E+00	−3.00E+00
	$\bar{f}$	−3.00E+00	7.96E−01	1.71E+00	−2.72E+00	−3.00E+00	−2.59E+00	−2.16E+00	−3.00E+00	−3.00E+00
	σ_f	6.40E−14	3.31E−01	1.01E+00	1.65E−01	5.82E−08	4.96E−01	3.98E−01	2.65E−05	1.62E−08
f_{14}	f_{Best}	2.75E+00	3.59E+01	3.34E+00	2.75E+00	2.75E+00	1.97E+01	2.75E+00	2.75E+00	**2.72E+00**
	f_{Worst}	1.34E−07	1.27E+01	8.86E−01	5.95E−06	9.84E−09	2.26E+01	1.05E−06	2.55E−04	2.75E+00
	$\bar{f}$	2.75E+00	1.49E+01	2.80E+00	2.75E+00	2.75E+00	2.75E+00	2.75E+00	2.75E+00	2.72E+00
	σ_f	2.75E+00	6.25E+01	7.68E+00	2.75E+00	2.75E+00	9.97E+01	2.75E+00	2.75E+00	5.59E−03
f_{15}	f_{Best}	9.96E−16	3.10E+00	1.13E+00	1.81E−06	2.97E−10	7.32E−10	4.76E−05	2.93E−10	**−1.04E+00**
	f_{Worst}	1.39E−14	7.40E+00	3.93E+00	1.03E−05	6.53E−08	4.00E+00	1.69E−04	3.03E−06	1.33E−14
	$\bar{f}$	4.59E−15	5.05E+00	2.21E+00	3.85E−06	6.86E−09	1.04E+00	7.72E−05	3.35E−07	−3.45E−02
	σ_f	3.41E−15	9.97E−01	7.48E−01	1.69E−06	1.22E−08	1.77E+00	2.54E−05	7.04E−07	1.89E−01
f_{16}	f_{Best}	**−1.25E+04**	−5.45E+03	−8.26E+03	−9.31E+03	−1.26E+04	−1.05E+04	−1.00E+04	−4.83E+03	**−1.25E+04**
	f_{Worst}	−1.20E+04	−3.73E+03	−4.14E+03	−8.38E+03	−1.19E+04	−7.20E+03	−5.56E+03	−3.52E+03	−3.83E+03
	$\bar{f}$	−1.22E+04	−4.40E+03	−5.81E+03	−8.65E+03	−1.23E+04	−8.68E+03	−7.92E+03	−4.07E+03	−4.95E+03

(continued)

Table 8.2 (continued)

		ABC	DE	PSO	CS	DS	MFO	MVO	SCA	EA-HC
	σ_f	1.19E+02	3.64E+02	9.44E+02	2.11E+02	1.91E+02	7.68E+02	8.60E+02	2.85E+02	2.12E+03
f_{17}	f_{Best}	8.06E−14	1.36E+06	1.31E+01	5.40E−01	1.09E−08	4.87E−07	7.55E−04	3.03E−01	**1.57E−32**
	f_{Worst}	2.12E−12	4.05E+07	3.42E+03	2.18E+00	1.04E−01	1.98E+00	4.59E+00	1.11E+01	1.12E−09
	$\overline{f}$	5.04E−13	1.64E+07	3.69E+02	1.15E+00	3.46E−03	2.93E−01	9.20E−01	1.20E+00	4.17E−11
	σ_f	3.97E−13	1.08E+07	9.05E+02	4.23E−01	1.89E−02	5.15E−01	1.00E+00	1.94E+00	2.05E−10
f_{18}	f_{Best}	5.99E−13	7.68E+06	6.98E+01	5.53E−02	2.36E−08	2.64E−05	7.68E−03	2.10E+00	**1.35E−32**
	f_{Worst}	1.33E−09	1.02E+08	2.83E+05	2.95E−01	9.49E−05	3.60E+00	1.21E−01	5.04E+00	1.56E−11
	$\overline{f}$	7.02E−11	4.89E+07	6.62E+04	1.59E−01	6.59E−06	2.03E−01	3.79E−02	2.73E+00	5.20E−13
	σ_f	2.41E−10	1.87E+07	7.98E+04	6.15E−02	1.72E−05	6.10E−01	2.22E−02	5.80E−01	2.84E−12
f_{19}	f_{Best}	2.67E−07	5.22E+03	1.21E+03	6.68E−01	2.12E−04	4.48E−04	3.09E−01	1.09E−06	**6.49E−81**
	f_{Worst}	1.70E−06	1.89E+04	2.23E+04	1.51E+01	2.98E−03	1.00E+05	7.33E−01	8.34E−03	4.34E+00
	$\overline{f}$	9.72E−07	1.35E+04	4.79E+03	3.64E+00	9.59E−04	2.53E+04	5.13E−01	1.77E−03	1.45E−01
	σ_f	4.09E−07	3.10E+03	4.07E+03	3.60E+00	7.20E−04	2.25E+04	1.22E−01	2.11E−01	7.93E−01
f_{20}	f_{Best}	**2.90E+01**	5.15E+02	2.33E+02	8.77E+01	**2.90E+01**	5.30E+01	9.13E+01	**2.90E+01**	**2.90E+01**
	f_{Worst}	2.90E+01	7.81E+02	4.66E+02	1.23E+02	4.73E+01	4.01E+02	1.98E+02	2.90E+01	3.04E+01
	$\overline{f}$	2.90E+01	6.30E+02	3.54E+02	1.09E+02	3.09E+01	1.29E+02	1.45E+02	2.90E+01	2.90E+01
	σ_f	1.00E−05	5.96E+01	5.38E+01	8.76E+00	4.78E+00	8.78E+01	2.45E+01	4.52E−03	2.59E−01
f_{21}	f_{Best}	2.78E+02	4.96E+07	1.12E+05	6.88E+01	1.56E+02	8.20E+06	5.13E+02	3.13E+02	**3.20E+01**
	f_{Worst}	3.26E+01	1.72E+07	1.64E+05	8.21E+01	4.96E+01	5.80E+07	6.52E+00	6.28E+02	3.07E+02
	$\overline{f}$	1.92E+02	2.10E+07	1.84E+03	5.93E+01	9.01E+01	4.79E+01	4.06E+01	3.20E+01	4.20E+01
	σ_f	3.38E+02	7.77E+07	6.82E+05	9.59E+01	2.63E+02	4.10E+08	6.80E+01	3.55E+03	5.02E+01

(continued)

Table 8.2 (continued)

		ABC	DE	PSO	CS	DS	MFO	MVO	SCA	EA-HC
f_{22}	f_{Best}	**2.90E+01**	5.45E+02	3.00E+02	1.27E+02	**2.90E+01**	4.28E+01	8.95E+01	**2.90E+01**	**2.90E+01**
	f_{Worst}	2.90E+01	9.86E+02	2.97E+03	1.84E+02	4.35E+01	2.53E+03	2.51E+02	2.90E+01	2.90E+01
	$\overline{f}$	2.90E+01	7.46E+02	7.10E+02	1.59E+02	2.99E+01	9.66E+02	1.66E+02	2.90E+01	2.90E+01
	σ_f	4.15E−06	1.07E+02	5.79E+02	1.50E+01	3.56E+00	5.61E+02	3.47E+01	3.79E−03	1.91E−04
f_{23}	f_{Best}	−8.38E+01	1.87E+06	1.40E+03	−7.29E+01	−8.31E+01	−8.27E+01	−8.17E+01	−3.23E+01	**−8.39E+01**
	f_{Worst}	−8.34E+01	2.82E+07	1.56E+06	−4.76E+01	−8.19E+01	1.02E+09	−1.30E+01	5.25E+02	−8.10E+01
	$\overline{f}$	−8.36E+01	9.86E+06	5.81E+04	−6.34E+01	−8.26E+01	1.84E+08	−7.68E+01	6.67E+00	−8.38E+01
	σ_f	1.17E−01	6.42E+06	2.83E+05	7.02E+00	3.69E−01	2.74E+08	1.65E+01	9.99E+01	5.39E−01

Table 8.3 Minimization results of benchmark functions of Table 8.12 with $n = 100$

		ABC	DE	PSO	CS	DS	MFO	MVO	SCA	EA–HC
f_1	f_{Best}	3.95E−03	8.61E+04	9.25E+03	1.34E+02	8.15E+01	2.27E+03	1.69E+01	2.23E+02	**1.05E−105**
	f_{Worst}	1.11E−01	1.18E+05	2.19E+04	3.10E+02	5.41E+02	5.01E+04	3.45E+01	1.25E+04	1.45E+01
	$\overline{f}$	2.25E−02	1.03E+05	1.55E+04	2.15E+02	1.95E+02	2.37E+04	2.32E+01	3.39E+03	4.83E−01
	σ_f	2.20E−02	6.98E+03	3.05E+03	4.36E+01	1.16E+02	1.28E+04	4.28E+00	3.17E+03	2.65E+00
f_2	f_{Best}	6.04E−02	1.35E+07	9.20E+01	1.00E+10	3.26E+00	8.46E+01	3.19E+02	9.23E−03	**2.15E−91**
	f_{Worst}	1.61E−01	2.82E+19	1.52E+02	1.00E+10	1.18E+01	3.33E+02	2.36E+16	6.47E+00	3.06E+00
	$\overline{f}$	9.62E−02	1.08E+18	1.17E+02	1.00E+10	6.60E+00	1.59E+02	1.43E+15	8.98E−01	1.74E−01
	σ_f	2.53E−02	5.15E+18	1.58E+01	0.00E+00	2.04E+00	4.98E+01	5.47E+15	1.36E+00	6.68E−01
f_3	f_{Best}	1.49E+05	3.82E+05	4.30E+04	3.32E+04	4.66E+04	1.02E+05	2.30E+04	9.75E+04	**6.89E−149**
	f_{Worst}	2.31E+05	7.80E+05	1.45E+05	6.17E+04	1.13E+05	3.09E+05	4.55E+04	2.49E+05	5.45E+04
	$\overline{f}$	1.85E+05	5.52E+05	7.65E+04	4.81E+04	7.19E+04	1.79E+05	3.15E+04	1.69E+05	7.16E+03
	σ_f	1.93E+04	9.79E+04	2.72E+04	7.64E+03	1.86E+04	5.21E+04	5.18E+03	4.46E+04	1.40E+04
f_4	f_{Best}	8.68E+01	8.16E+01	2.81E+01	1.38E+01	2.85E+01	8.80E+01	3.11E+01	7.66E+01	**3.01E−133**
	f_{Worst}	9.26E+01	8.90E+01	4.38E+01	2.16E+01	5.53E+01	9.57E+01	5.64E+01	9.11E+01	2.33E−51
	$\overline{f}$	9.01E+01	8.53E+01	3.61E+01	1.72E+01	4.34E+01	9.22E+01	4.23E+01	8.50E+01	8.77E−53
	σ_f	1.72E+00	1.52E+00	3.95E+00	2.01E+00	6.01E+00	1.96E+00	6.57E+00	3.29E+00	4.27E−52
f_5	f_{Best}	9.66E+01	1.26E+03	3.29E+03	9.96E+01	9.73E+01	2.86E+02	9.16E+01	1.21E+02	**1.75E−01**
	f_{Worst}	1.18E+02	1.70E+03	6.04E+03	1.03E+02	1.64E+02	6.98E+02	1.86E+02	4.16E+02	9.80E+01
	$\overline{f}$	1.04E+02	1.44E+03	4.62E+03	1.01E+02	1.16E+02	4.86E+02	1.05E+02	2.34E+02	7.43E+01
	σ_f	6.05E+00	1.11E+02	6.65E+02	1.06E+00	1.97E+01	1.13E+02	2.24E+01	8.39E+01	4.02E+01
f_6	f_{Best}	4.00E+00	8.90E+04	1.15E+04	3.37E+02	1.24E+02	3.96E+03	8.50E+01	4.81E+02	**0.00E+00**

(continued)

Table 8.3 (continued)

f_7	f_{Worst}	2.80E+01	1.19E+05	2.14E+04	5.87E+02	9.14E+02	4.51E+04	2.34E+02	1.18E-04	0.00E+00
	$\bar{f}$	1.28E+01	1.03E+05	1.64E+04	4.48E+02	3.20E+02	2.77E+04	1.53E-02	4.20E-03	0.00E+00
	σ_f	6.09E+00	7.81E+03	2.41E+03	7.26E+01	1.87E+02	1.06E+04	4.12E-01	3.19E+03	0.00E+00
f_8	f_{Best}	1.38E+00	2.32E+02	3.19E+02	3.25E-01	8.16E-01	4.70E+00	1.36E-01	4.70E+00	**6.98E−06**
	f_{Worst}	2.79E+00	4.55E+02	7.56E+02	8.35E-01	2.35E+00	4.16E+02	3.88E-01	1.13E-02	2.52E-03
	$\bar{f}$	2.26E+00	3.58E+02	5.01E+02	5.14E-01	1.30E+00	1.41E+02	2.43E-01	4.86E+01	3.19E-04
	σ_f	3.25E-01	5.24E+01	1.01E+02	1.22E-01	4.04E-01	1.04E+02	6.36E-02	3.03E+01	5.18E-04
f_9	f_{Best}	2.03E+00	5.72E+01	3.12E+01	5.19E+00	2.03E+00	2.04E+00	2.03E+00	2.11E+00	**2.02E+00**
	f_{Worst}	2.10E+00	6.69E+01	4.47E+01	8.87E+00	2.04E+00	6.38E+00	2.88E+00	2.29E+00	3.69E-01
	$\bar{f}$	2.05E+00	6.22E+01	3.77E+01	7.03E+00	2.03E+00	3.18E+00	2.23E+00	2.18E+00	3.53E+00
	σ_f	1.47E-02	2.40E+00	3.37E+00	8.48E-01	3.31E-03	1.35E+00	2.10E-01	3.57E-02	6.42E+00
f_{10}	f_{Best}	2.34E-02	4.85E+01	4.05E+01	1.59E-01	3.67E-02	4.54E-02	2.40E-02	9.74E-02	**2.04E−02**
	f_{Worst}	3.65E-02	6.56E+01	5.90E+01	4.77E-01	1.15E-01	8.11E+00	2.64E-02	4.80E-01	7.35E+00
	$\bar{f}$	2.68E-02	5.66E+01	5.00E+01	3.20E-01	7.09E-02	1.70E+00	2.53E-02	1.87E-01	4.99E-01
	σ_f	3.05E-03	3.96E+00	4.03E+00	9.18E-02	2.33E-02	2.75E+00	4.96E-04	7.26E-02	1.82E+00
f_{11}	f_{Best}	4.38E+01	1.14E+03	6.49E+02	4.08E+02	1.62E+02	5.76E+02	4.43E-02	1.05E-01	**0.00E+00**
	f_{Worst}	8.54E+01	1.27E+03	9.56E+02	5.41E+02	3.54E+02	8.57E+02	7.29E+02	5.13E+02	9.43E+02
	$\bar{f}$	6.60E+01	1.21E+03	8.19E+02	4.73E+02	2.31E+02	7.12E+02	5.91E+02	2.08E+02	4.64E+01
	σ_f	1.06E+01	2.88E+01	7.23E+01	3.38E+01	4.46E+01	6.04E+01	7.69E+01	1.05E+02	1.77E+02
	f_{Best}	2.25E+00	1.88E+01	1.21E+01	6.63E+00	3.28E+00	1.90E+01	2.91E+00	6.52E-02	**8.88E−16**
	f_{Worst}	3.73E+00	1.96E+01	1.46E+01	1.70E+01	1.99E+01	2.00E+01	1.99E+01	2.06E+01	4.85E-07
	$\bar{f}$	3.03E+00	1.93E+01	1.34E+01	1.09E+01	5.22E+01	1.98E+01	5.23E+00	1.85E+01	1.99E-08

(continued)

Table 8.3 (continued)

f_{12}	σ_f	4.06E−01	1.81E−01	6.22E−01	2.67E+00	3.13E+00	2.71E−01	4.98E+00	4.63E+00	8.93E−08
	f_{Best}	1.74E−02	7.90E+02	8.74E+01	2.26E+00	1.64E+00	4.26E+01	1.14E+01	6.96E+00	**0.00E+00**
	f_{Worst}	5.92E−01	1.03E+03	1.89E+02	3.95E+00	9.16E+00	5.65E+02	1.24E+00	1.31E+02	1.02E−09
	$\overline{f}$	1.96E−01	9.16E+02	1.41E+02	2.94E+00	3.05E+00	2.32E+02	1.20E+00	4.14E+01	3.39E−11
f_{13}	σ_f	1.55E−01	6.06E+01	2.70E+01	4.24E−01	1.77E+00	1.27E+02	2.17E−02	3.45E+01	1.85E−10
	f_{Best}	**−1.00E+01**	6.69E+00	1.64E+01	−8.65E+00	−9.54E+00	−6.52E+00	−6.72E+00	−9.97E+00	**−1.00E+01**
	f_{Worst}	−9.70E+00	1.13E+01	2.66E+01	−6.25E+00	−7.32E+00	−6.82E−01	−3.32E+00	−7.39E+00	−1.00E+01
	$\overline{f}$	−9.90E+00	9.86E+00	2.09E+01	−7.23E+00	−8.63E+00	−3.64E+00	−5.23E+00	−9.09E+00	−1.00E+01
f_{14}	σ_f	9.79E−02	9.98E−01	2.64E+00	5.83E−01	5.02E−01	1.54E+00	8.49E−01	8.30E−01	8.66E−10
	f_{Best}	2.73E+00	6.02E+02	1.50E+01	3.22E+00	2.76E+00	6.05E+01	2.73E+00	5.19E+00	**2.72E+00**
	f_{Worst}	5.48E+00	1.23E+03	5.94E+01	4.13E+00	5.49E+00	6.84E+02	2.73E+00	2.42E+02	2.73E+00
	$\overline{f}$	3.40E+00	8.62E+02	2.50E+01	3.69E+00	3.18E+00	2.33E+02	2.73E+00	5.77E+01	2.72E+00
f_{15}	σ_f	6.22E−01	1.42E+02	9.54E+00	2.70E−01	5.71E−01	1.46E+02	1.41E−05	6.60E+01	3.34E−03
	f_{Best}	1.67E−04	3.23E+01	1.76E+01	5.88E−02	3.03E−02	8.47E−01	5.81E−03	1.11E−01	**1.06E−165**
	f_{Worst}	4.82E−03	4.67E+01	3.00E+01	1.34E−01	2.44E−01	2.06E+01	1.13E−02	5.85E+00	1.13E−01
	$\overline{f}$	9.76E−04	3.94E+01	2.37E+01	8.51E−02	8.08E−02	1.08E+01	8.66E−03	1.47E+00	3.76E−03
f_{16}	σ_f	9.26E−04	3.14E+00	3.72E+00	1.96E−02	4.47E−02	5.57E+00	1.29E−03	1.42E+00	2.06E−02
	f_{Best}	−3.64E+04	−8.99E+03	−9.27E+03	−2.33E+04	−3.14E+04	−2.90E+04	−2.74E+04	−8.19E+03	**−4.19E−04**
	f_{Worst}	−3.43E+04	−6.61E+03	−5.72E+03	−2.11E+04	−2.74E+04	−2.03E+04	−2.20E+04	−6.58E+03	−7.35E+03
	$\overline{f}$	−3.52E+04	−7.86E+03	−7.61E+03	−2.19E+04	−2.91E+04	−2.38E+04	−2.44E+04	−7.34E+03	−1.81E+04
f_{17}	σ_f	6.17E+02	6.87E+02	1.05E+03	4.63E+02	1.06E+03	2.62E+03	1.56E+03	3.99E+02	1.47E+04
	f_{Best}	3.28E−05	2.47E+08	2.58E+04	4.40E+00	2.74E+00	3.93E+04	2.58E+04	2.51E+07	**4.71E−33**

(continued)

Table 8.3 (continued)

f_{18}	f_{Worst}	3.11E−02	6.70E+08	9.29E+05	8.46E+00	1.42E+01	5.22E+08	9.29E+05	3.11E+08	4.20E−09
	$\bar{f}$	1.26E−03	4.48E+08	2.84E+05	5.67E+00	6.58E+00	1.08E+08	2.84E+05	1.16E+08	1.40E−10
	σ_f	5.65E−03	9.17E+07	2.72E+05	7.80E−01	2.43E+00	1.61E+08	2.72E+05	6.55E+07	7.66E−10
f_{19}	f_{Best}	5.75E−04	6.37E+08	1.89E+06	6.13E+01	4.02E+01	9.25E+06	1.89E+06	4.98E+07	**1.35E−32**
	f_{Worst}	5.15E−02	1.21E+09	2.22E+07	5.88E+02	3.17E+02	4.57E+08	2.22E+07	5.70E+08	3.71E−12
	$\bar{f}$	4.43E−03	9.42E+08	7.18E+06	1.27E+02	8.65E+01	1.89E+08	7.18E+06	2.48E+08	1.49E−13
	σ_f	9.47E−03	1.64E+08	4.60E+06	9.95E+01	4.93E+01	1.80E+08	4.60E+06	1.23E+08	6.80E−13
f_{20}	f_{Best}	3.82E−02	3.53E+12	1.44E+04	1.00E+10	7.59E+01	7.40E+04	1.44E+04	2.14E+02	**2.00E−104**
	f_{Worst}	1.83E−01	1.12E+20	6.63E+04	1.00E+10	4.75E+02	2.32E+05	6.63E+04	1.23E+04	2.88E+00
	$\bar{f}$	7.61E−02	9.05E+18	2.20E+04	1.00E+10	2.14E+02	1.37E+05	2.20E+04	4.57E+03	9.60E−02
	σ_f	2.84E−02	2.48E+19	9.22E+03	0.00E+00	9.09E+01	4.78E+04	9.22E+03	3.31E+03	5.25E−01
f_{21}	f_{Best}	1.47E+02	3.06E+03	1.27E+03	5.34E+02	3.25E+02	9.81E+02	1.27E+03	1.41E+02	**9.90E+01**
	f_{Worst}	2.12E+02	4.18E+03	1.99E+03	6.61E+02	5.18E+02	3.04E+03	1.99E+03	9.59E+02	9.90E+01
	$\bar{f}$	1.81E+02	3.66E+03	1.44E+03	6.01E+02	4.03E+02	1.86E+03	1.44E+03	5.03E+02	9.90E+01
	σ_f	1.70E+01	2.56E+02	1.39E+02	2.73E+01	4.79E+01	5.34E+02	1.39E+02	2.06E+02	4.40E−04
f_{22}	f_{Best}	2.81E+03	7.38E+08	2.03E+06	2.41E+03	6.78E+03	6.63E+06	2.03E+06	3.61E+07	**1.09E+02**
	f_{Worst}	5.74E+03	1.33E+09	1.83E+07	7.10E+03	1.16E+05	1.27E+09	1.83E+07	7.69E+08	3.10E+03
	$\bar{f}$	3.86E+03	1.00E+09	6.53E+06	3.78E+03	1.72E+04	2.25E+08	6.53E+06	2.56E+08	3.92E+02
	σ_f	6.81E+02	1.48E+08	4.25E+06	8.91E+02	1.97E+04	3.10E+08	4.25E+06	1.47E+08	7.86E+02
	f_{Best}	9.95E+01	5.62E+09	1.43E+03	1.00E+10	2.34E+02	1.88E+03	1.43E+03	1.26E+02	**9.90E+01**
	f_{Worst}	1.54E+02	8.00E+18	2.92E+03	1.00E+10	4.27E+02	7.43E+03	2.92E+03	1.02E+03	9.90E+01
	$\bar{f}$	1.13E+02	3.12E+17	1.70E+03	1.00E+10	3.15E+02	4.48E+03	1.70E+03	3.66E+02	9.90E+01

(continued)

Table 8.3 (continued)

f_{23}	σ_f	1.40E+01	1.47E+18	3.07E+02	0.00E+00	4.54E+01	1.23E+03	3.07E+02	2.31E+02	4.02E−03
	f_{Best}	−2.95E+02	1.82E+10	1.62E+04	1.00E+10	−1.24E+02	2.66E+08	1.62E+04	4.90E+05	**−2.98E+02**
	f_{Worst}	−2.91E+02	7.53E+20	2.23E+06	1.00E+10	1.37E+03	6.40E+09	2.23E+06	1.43E+08	−2.87E+02
	$\overline{f}$	−2.94E+02	3.22E+19	4.53E+05	1.00E+10	5.15E+02	2.81E+09	4.53E+05	2.67E+07	−2.97E+02
	σ_f	7.66E−01	1.38E+20	5.90E+05	0.00E+00	4.13E+02	1.52E+09	5.90E+05	3.13E+07	1.88E+00

competes directly with the results of EA-HC in function f_{16}. Finally, the ABC, DS, and SCA algorithms obtain the same results as the proposed algorithm.

By the numerical results of Table 8.2, it can be demonstrated that EA-HC obtains a better response against the compared metaheuristic approaches for the majority of benchmark functions. Only a few reported results suggest that the performance of the proposed method achieves similar results than some metaheuristic algorithms. The remarkable performance of the proposed method is based on its capability of changing the optimization process between exploration and exploitation stages. Traditionally, the metaheuristic operators of metaheuristic optimization algorithms are design to start performing exploration in the search space; then, at the final stage of the optimization process, the exploitation mechanism is performed. In the proposed mechanism, a changing frequency is added to perform both exploration and exploitation during the entire optimization process. This mechanism improves the search strategy towards the global optimum by changing the metaheuristic stages. By this changing scheme, the proposed method is capable of obtaining the best results for most of the benchmark functions evaluating a 30-dimensional search space.

To test the scalability of the proposed method in higher-dimensional search spaces, a performance comparison among EA-HC and the rest of tested metaheuristic methods considering a 100-dimensional search is conducted. Table 8.3 reports the numerical results for this test. According to the table, it can be deduced that the metaheuristic operators of the proposed approach outperform the rest of its competitors evaluating higher-dimensional search spaces. The structure of the metaheuristic operators of EA-HC, conducts the search strategy into a more efficient mechanism than the rest of the algorithms evaluating most of the benchmark functions in Table 8.12. An exception to this is in function f_{13}. Under the numerical results, is quite evident that only the ABC method performs quite similar than EA-HC. If the mean and standard deviation are analyzed then, the ABC produces more consistent results than EA-HC. This may indicate that the proposed method produces less consistent results. However, the main reason for this phenomenon is based on the changing character of the proposed method. EA-HC method uses the execution of the exploration and exploitation operators during the entire optimization process in counterpart than most of the metaheuristic methodologies which incorporate an exploration and exploitation stages in a fixed period of time in the optimization process. The main advantage of EA-HC over the rest of the tested algorithms is achieved by the balance of the evolutionary stages over the entire optimization process. As a principle, EA-HC considers a changing scheme among exploration and exploitation every 50 iterations; this mechanism allows EA-HC to produce a balance among evolutionary stages which conducts the search strategy by maintaining the population diversity.

As a result, it is evident that the proposed method presents a remarkable performance in higher-dimensional search spaces. The structure of the evolutionary mechanism of EA-HC, produces better results than its competitors. The numerical results in higher dimensional search spaces suggest that the changing character of EA-HC presents a higher level of scalability since it outperforms the rest of tested algorithms in the majority of benchmark functions.

To statistically corroborate the numerical results from Tables 8.2 and 8.3, a non-parametric test is conducted. In this study, the Wilcoxon rank-sum test [44] has been conducted in order to validate the performance results. The Wilcoxon test has been applied considering the 0.05 significance value over the 30 independent runs for each benchmark function. Table 8.4 reports the p-values considering the performance results of Table 8.2 (where $n = 30$) obtained by the Wilcoxon test. Additionally, Table 8.5 reports the p-values considering the performance results of Table 8.3 (where $n = 100$) obtained by the rank-sum test. In Tables 8.4 and 8.5, a pairwise comparison among EA-HC and the rest of the tested algorithms is presented. Under such circumstances, eight groups are considered in the study; EA-HC versus ABC, EA-HC versus DE, EA-HC versus PSO, EA-HC versus CS, EA-HC versus DS, EA-HC versus MFO, EA-HC versus MVO and EA-HC versus SCA.

In the statistical experiment, it is considered that there is no significant difference in a group (null hypothesis H_0). Also, it is considered as an alternative hypothesis H_1 that there is a significant difference in a group. To make a clear visualization of the results, Tables 8.4 and 8.5, adopt the symbols▲ ▼, and ►. The symbol ▲, represents that EA-HC achieves significantly better results than a given competitor. The symbol ▼, represents that EA-HC produces worse results than its competitor, and the symbol ► represents the situation when the Wilcoxon test is not able to distinguish between the numerical results. According to the p-values from Table 8.4, it is demonstrated that for function f_6 the groups EA-HC versus ABC, EA-HC versus CS, EA-HC versus DS, EA-HC versus MFO and EA-HC versus SCA the rank-sum is not able to distinguish among the results. For function f_8, the groups EA-HC versus ABC, EA-HC versus CS and EA-HC versus DS corroborates that the proposed method and these methods produce quite similar results. According to the Wilcoxon test, the p-values for functions f_9 and f_{13}, suggest that EA-HC produces similar results than ABC, CS, DS, MFO and SCA. In function f_{16}, the p-value from the group EA-HC versus ABC indicates that the rank-sum is not able to distinguish among the numerical results from both algorithms. Finally, the groups EA-HC versus ABC, EA-HC versus DS and EA-HC versus SCA, statistically corroborates the results from Table 8.4 in function f_{20}.

According to the p-values from Table 8.5, it is quite evident the superior performance of the proposed approach against each metaheuristic algorithm considered in the experimental study. As it can be demonstrated, only in function f_{13}, the Wilcoxon test indicates that EA-HC and ABC perform quite similar. For the rest of the entries, the EA-HC method outperforms the rest of the competitors considering 100-dimensional search spaces.

From the statistical results reported in Tables 8.4 and 8.5, it can be deduced that the changing character of the evolutionary structure of the proposed approach produces more robust and scalable results than the rest of metaheuristic algorithms maintaining the population diversity over the optimization process.

Table 8.4 P-values produced by Wilcoxon rank sum test comparing EA-HC versus ABC, EA-HC versus DE, EA-HC versus PSO, EA-HC versus CS, EA-HC versus DS, EA-HC versus MFO, EA-HC versus MVO and EA-HC versus SCA over the averaged fitness value $\overline{f}$ for each function from Table 8.2

EA-HC vs	ABC	DE	PSO	CS	DS	MFO	MVO	SCA
f_1	1.066E−07▲	1.51E−08▲	2.2498E−07▲	6.7650E−05▲	1.4918E−06▲	8.4766E−05▲	6.7650E−05▲	1.6351E−05▲
f_2	9.9584E−05▲	3.0199E−11▲	3.0199E−11▲	3.0103E−07▲	3.0376E−06▲	3.0610E−E−11▲	1.1937E−06▲	6.4632E−10▲
f_3	6.1846E−10▲	3.0199E−11▲	5.4620E−06▲	3.0199E−11▲	3.0199E−11▲	5.1836E−10▲	3.0199E−11▲	7.3803E−10▲
f_4	6.0459E−07▲	3.0199E−11▲	1.1737E−09▲	3.0199E−11▲	3.0199E−11▲	2.5346E−12▲	3.0199E−11▲	3.3384E−11▲
f_5	3.0199E−11▲	3.0199E−11▲	3.6897E−11▲	3.0199E−11▲	6.6955E−11▲	1.3578E−07▲	1.6132E−10▲	3.0199E−11▲
f_6	0.0000E00▲	4.6457E−08▲	2.7013E−03▲	0.0000E00▲	0.0000E00▲	0.0000E00▲	1.8255E−04▲	0.0000E00▶
f_7	3.0199E−11▲	6.0658E−11▲	7.0430E−07▲	3.0199E−11▲	3.0199E−11▲	1.3744E−05▲	3.0199E−11▲	3.0199E−11▲
f_8	0.0000E00▲	2.3885E−04▲	5.5611E−04▲	0.0000E00▲	0.0000E00▶	9.4249E−14▲	0.0000E00▶	3.3384E−11▲
f_9	0.0000E00▲	3.0199E−11▲	3.0199E−11▲	0.0000E00▶	0.0000E00▶	9.4342E−14▲	3.0199E−11▲	3.0199E−11▲
f_{10}	4.9752E−11▲	6.5183E−09▲	5.4342E−08▲	3.2798E−02▲	3.4971E−09▲	4.3422E−05▲	7.3412E−07▲	7.6950E−08▲
f_{11}	6.7493E−05▲	1.4229E−08▲	1.3357E−02▲	3.9807E−04▲	6.7493E−05▲	4.2628E−04▲	3.9807E−06▲	3.5597E−06▲
f_{12}	4.9980E−09▲	1.4733E−07▲	1.1279E−05▲	1.1077E−06▲	1.1567E−07▲	1.199E−05▲	1.1077E−06▲	4.1127E−07▲
f_{13}	0.0000E00▲	2.2623E−07▲	2.5040E−11▲	3.0199E−11▲	0.0000E00▶	0.0000E00▶	1.0425E−07▲	0.0000E00▶
f_{14}	3.0199E−11▲	1.7769E−10▲	4.0772E−11▲	3.0199E−11▲	3.0199E−11▲	1.9643E−11▲	3.0199E−11▲	3.0199E−11▲
f_{15}	5.5727E−10▲	1.0105E−08▲	6.8762E−09▲	1.0666E−07▲	8.4848E−09▲	5.1399E−06▲	1.1567E−07▲	1.4294E−08▲
f_{16}	0.0000E00▶	3.2310E−05▲	6.7362E−06▲	3.0199E−11▲	3.0199E−11▲	9.4342E−14▲	4.0772E−11▲	1.3594E−07▲
f_{17}	3.0199E−11▲	3.0199E−11▲	6.1210E−10▲	3.1589E−10▲	3.0199E−11▲	1.5945E−13▲	1.2057E−10▲	1.2057E−10▲
f_{18}	3.0199E−11▲	8.1527E−11▲	5.5727E−10▲	3.0199E−11▲	3.0199E−11▲	1.0172E−13▲	3.0199E−11▲	1.2057E−11▲
f_{19}	1.1077E−06▲	2.2273E−09▲	1.6216E−05▲	6.7650E−05▲	1.1077E−06▲	4.9135E−05▲	6.7650E−05▲	1.4918E−06▲
f_{20}	0.0000E00▶	2.1544E−10▲	5.4974E−06▲	8.4848E−09▲	0.0000E00▶	4.6707E−09▲	7.6950E−08▲	0.0000E00▶

(continued)

Table 8.4 (continued)

EA-HC vs	ABC	DE	PSO	CS	DS	MFO	MVO	SCA
f_{21}	3.0199E−11 ◄	3.0199E−11 ◄	3.6897E−11 ◄	3.0199E−11 ◄	3.0199E−11 ◄	8.6495E−13 ◄	3.0199E−11 ◄	3.0199E−11 ◄
f_{22}	1.1073E−06 ◄	5.4907E−11 ◄	3.8741E−03 ◄	2.5967E−05 ◄	1.1933E−06 ◄	2.5646E−07 ◄	3.1565E−05 ◄	1.1073E−06 ◄
f_{23}	3.0199E−11 ◄	3.099E−11 ◄	5.6073E−05 ◄	3.0199E−11 ◄	3.0199E−11 ◄	9.9864E−09 ◄	3.0199E−11 ◄	6.7220E−10 ◄
◄	17	23	23	20	18	21	22	20
►	0	0	0	0	0	0	0	0
▲	6	0	4	3	5	2	1	3

Table 8.5 P-values produced by Wilcoxon rank sum test comparing EA-HC versus ABC, EA-HC versus DE, EA-HC versus PSO, EA-HC versus CS, EA-HC versus DS, EA-HC versus MFO, EA-HC versus MVO and EA-HC versus SCA over the averaged fitness value $\bar{f}$ for each function from Table 8.3

EA-HC vs	ABC	DE	PSO	CS	DS	MFO	MVO	SCA
f_1	3.0199E−11▲	3.0199E−11▲	3.0199E−11▲	3.0199E−11▲	3.0199E−11▲	2.1947E−08▲	3.0199E−11▲	3.0199E−11▲
f_2	1.0666E−07▲	3.0199E−11▲	2.5721E−07▲	1.2118E−12▲	2.5306E−04▲	5.4617E−09▲	3.0199E−11▲	6.5261E−07▲
f_3	3.0811E−08▲	3.0199E−11▲	1.1937E−06▲	3.0199E−11▲	3.6459E−08▲	1.6813E−04▲	3.0199E−11▲	8.1200E−04▲
f_4	3.0199E−11▲	3.0199E−11▲	6.2828E−06▲	3.0199E−11▲	3.0199E−11▲	3.0199E−11▲	6.7362E−06▲	3.0199E−11▲
f_5	3.0199E−11▲	3.0199E−11▲	3.0199E−11▲	3.0199E−11▲	3.0199E−11▲	7.2313E−09▲	3.0199E−11▲	3.0199E−11▲
f_6	2.9822E−11▲	3.0199E−11▲	3.0199E−11▲	3.0142E−11▲	3.0199E−11▲	1.3594E−07▲	3.0180E−11▲	3.0199E−11▲
f_7	3.0199E−11▲	3.0199E−11▲	3.0199E−11▲	3.0199E−11▲	3.0199E−11▲	6.2027E−04▲	3.0199E−11▲	4.3298E−05▲
f_8	3.0199E−11▲	2.1386E−06▲	8.3520E−08▲	3.0199E−11▲	3.0199E−11▲	3.0199E−11▲	3.0199E−11▲	3.0199E−11▲
f_9	3.0199E−11▲	2.0283E−07▲	4.1825E−09▲	3.0199E−11▲	3.0199E−11▲	3.0199E−11▲	3.0199E−11▲	3.0199E−11▲
f_{10}	3.0199E−11▲	3.0199E−11▲	7.0646E−09▲	3.6524E−11▲	4.4440E−07▲	9.5243E−08▲	4.1862E−05▲	6.5261E−07▲
f_{11}	5.5727E−10▲	3.0199E−11▲	6.1210E−10▲	8.4848E−09▲	7.7725E−09▲	3.0199E−11▲	8.8411E−07▲	6.7362E−06▲
f_{12}	3.0199E−11▲	3.0199E−11▲	2.3715E−10▲	3.0199E−11▲	3.0199E−11▲	2.1959E−07▲	3.0199E−11▲	3.3384E−11▲
f_{13}	3.0199E−11▲	3.0199E−11▲	3.0199E−11▲	5.5727E−10▲	3.8202E−10▲	3.0939E−06▲	2.2273E−09▲	9.9186E−11▲
f_{14}	3.0199E−11▲	3.0199E−11▲	5.4941E−11▲	3.0199E−11▲	3.0199E−11▲	1.2493E−05▲	3.0199E−11▲	1.5292E−05▲
f_{15}	5.5727E−10▲	3.0199E−11▲	3.8307E−05▲	5.5727E−10▲	5.5727E−10▲	1.0188E−05▲	5.5727E−10▲	8.8910E−10▲
f_{16}	3.0199E−11▲	1.1058E−04▲	1.0407E−04▲	3.0199E−11▲	3.0199E−11▲	3.0199E−11▲	3.0199E−11▲	4.1825E−09▲
f_{17}	3.0199E−11▲	3.0199E−11▲	2.0152E−08▲	3.0199E−11▲	3.0199E−11▲	4.9818E−04▲	3.0199E−11▲	3.0199E−11▲
f_{18}	3.0199E−11▲	3.0199E−11▲	3.6897E−11▲	3.0199E−11▲	3.0199E−11▲	2.3452E−10▲	3.0199E−11▲	2.6099E−10▲
f_{19}	3.0199E−11▲	3.0199E−11▲	2.1947E−08▲	1.2118E−12▲	3.0199E−11▲	3.0199E−11▲	3.0199E−11▲	3.0199E−11▲
f_{20}	3.0199E−11▲	3.0199E−11▲	2.6695E−09▲	3.0199E−11▲	3.0199E−11▲	1.3546E−06▲	3.3384E−06▲	3.0199E−11▲

(continued)

Table 8.5 (continued)

EA-HC vs	ABC	DE	PSO	CS	DS	MFO	MVO	SCA
f_{21}	3.0199E−11◄	3.0199E−11◄	1.2870E−09◄	3.0199E−11◄	3.0199E−11◄	9.7538E−05◄	3.0199E−11◄	8.1014E−10◄
f_{22}	3.0199E−11◄	3.0199E−11◄	5.8365 E−11◄	1.2118E−12◄	3.8202E−10◄	7.3891E−11◄	3.0199E−11◄	1.6132E−10◄
f_{23}	0.0000E00►	3.0199E−11◄	1.3289E−10◄	1.2118E−12◄	3.0199E−11◄	3.0199E−11◄	5.5727E−10◄	1.5846E−06◄
◄	22	23	23	23	23	23	23	23
►	0	0	0	0	0	0	0	0
▲	1	0	0	0	0	0	0	0

8.5.3 Convergence

In this section, convergence analysis of the proposed approach and the tested meta-heuristic algorithms is presented. The performance comparison exposed in Tables 8.2 and 8.3, reports the capabilities of the proposed approach in terms of fitness values. However, in most of the reported literature, a converge study must be included to evaluate the velocity which metaheuristic approaches reach during the optimization process for each benchmark function. In the convergence experiment, the convergence data was selected based on the average fitness values and considering the evaluation over 100-dimensional search spaces from Table 8.3.

Figure 8.4 indicates that the convergence rate of the proposed method is the fastest regarding the convergence speed of the tested algorithms. As a result, the remarkable performance of EA-HC, suggests that its changing characteristic produces more reliable results by maintaining the population diversity during the entire optimization process.

8.5.4 Engineering Design Problems

Optimization problems are defined as a mathematical model to represent many real-world problems. The main purpose of optimization is the finding process of the optimal solution for a given objective function. Since many disciplines, such as engineering, medicine, economics, etc. formulate their problems in terms of minimization/maximization, recently developed metaheuristic techniques should be tested under such examples. Traditionally, to measure the performance on meta-heuristic algorithms over real-world applications, several engineering design problems are evaluated as objective functions. In this study, the performance of EA-HC is tested over three common engineering design problems [21, 48]; The three-bar truss design problem (Table 8.13 in Appendix 8.2), the tension/compression spring design problem (Table 8.14 in Appendix 8.2), and the welded beam design problem (Table 8.15 in Appendix 8.2). The following section presents the numerical results for this test.

8.5.4.1 Three-Bar Truss Design Problem

The three-bar truss design optimization is conceived as a design optimization problem design. The main purpose of this task is to minimize the volume of a loaded three-bar truss, which has been subjected to several design constraints by each truss. The mathematical model of this optimization task considers a 2-dimensional search space with three inequalities. The experimental setting considers the configuration described in Sect. 8.5.1 with $gen = 1000$ for each algorithm. The graphical description of the problem is presented in Fig. 8.5, and the numerical results are presented in Table 8.6.

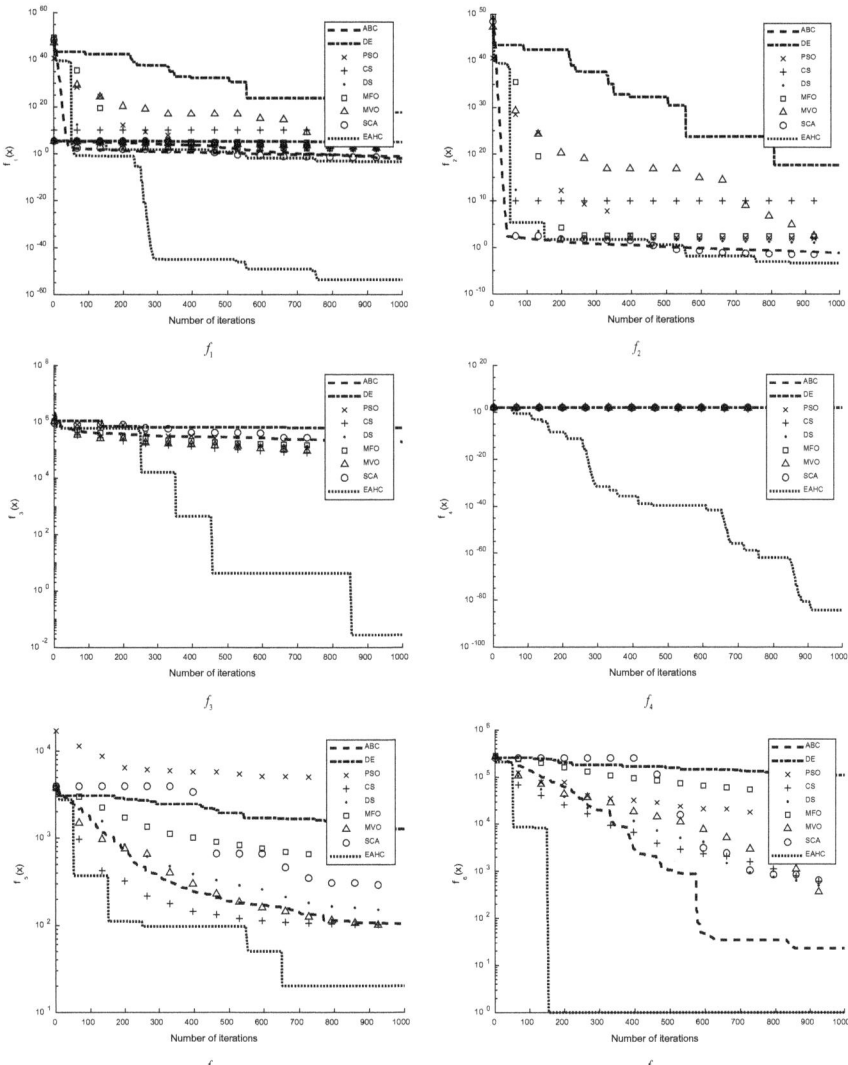

Fig. 8.4 Convergence graphs from Table 8.3

In the table, the parameters of the optimization problem are presented. Such entries indicate the decision variables, the constraints and the fitness value.

Table 8.6 indicates that the proposed method is capable of obtaining competitive results than CS, DS and MVO methods. To statistically validate the results from Tables 8.6 and 8.7 reports the worst, mean, standard deviation and the best fitness values achieved by each metaheuristic technique. As it can be deduced, the proposed EA-HC achieves similar fitness value than CS, DS and MVO. However, it presents

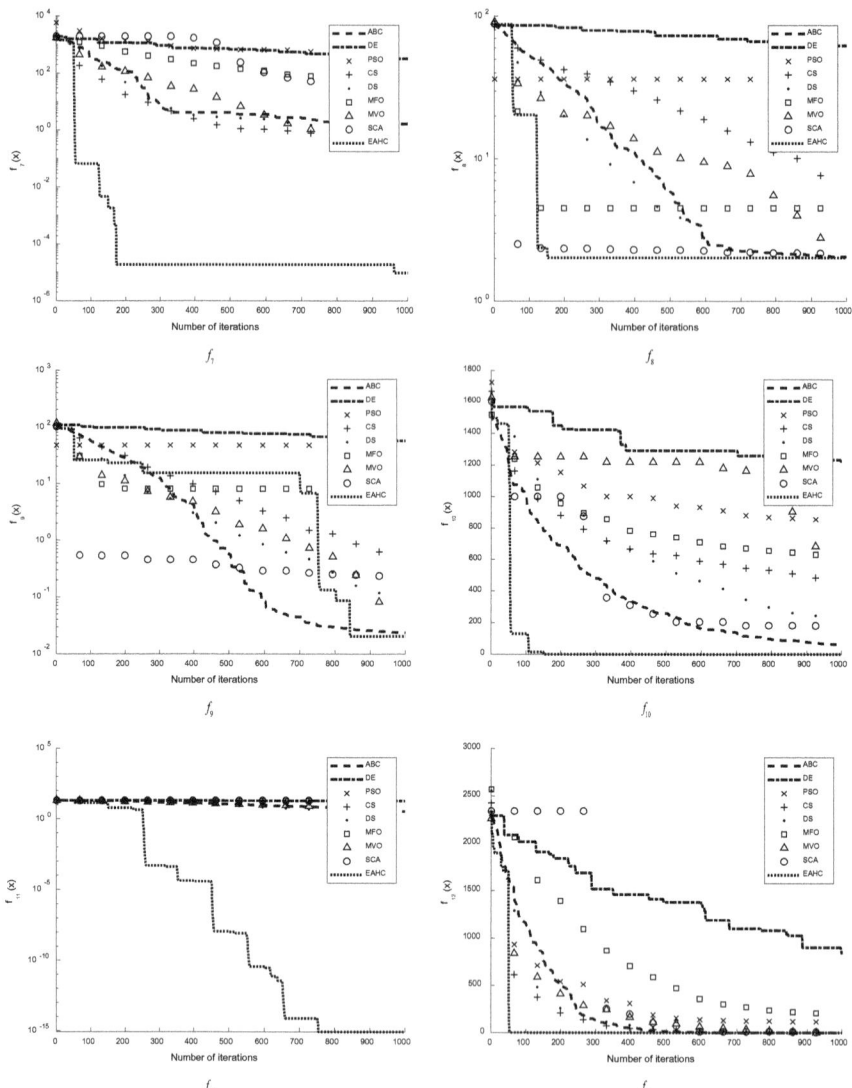

Fig. 8.4 (continued)

higher mean and standard deviation values due to its changing characteristics in the evolutionary structure.

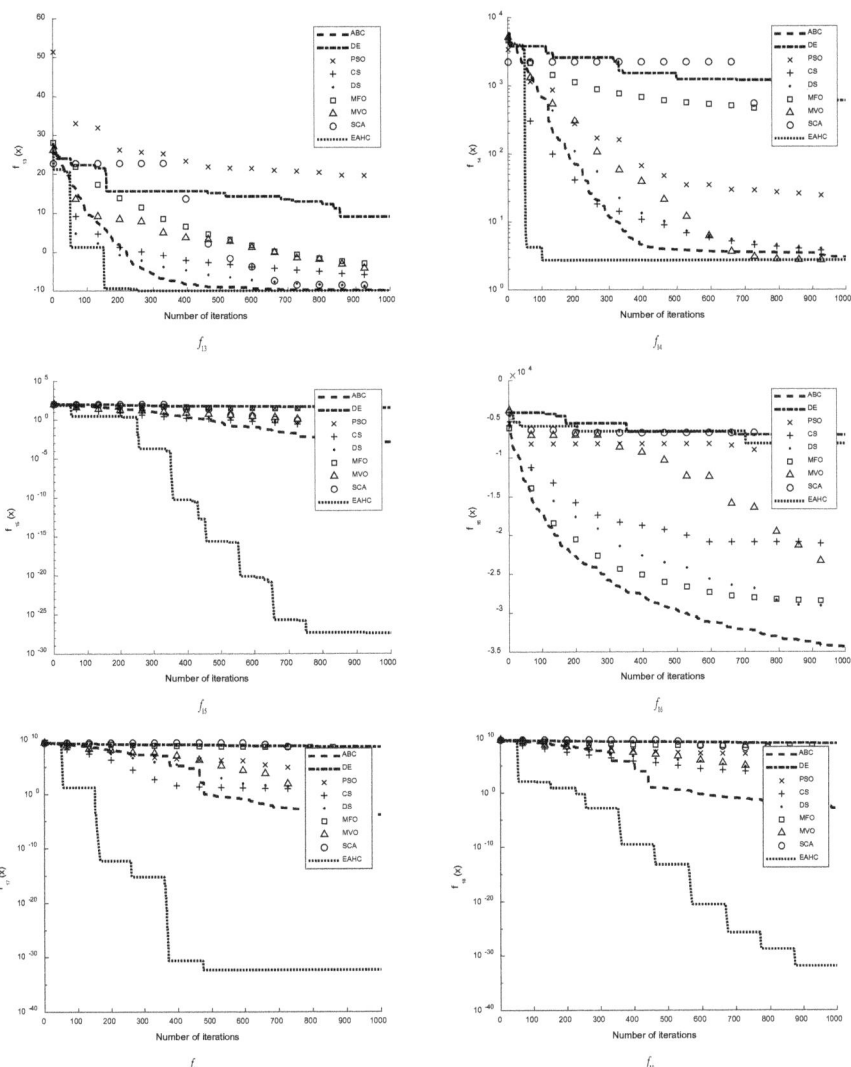

Fig. 8.4 (continued)

8.5.4.2   Tension/Compression Spring Design Problem

The spring design optimization task presents an optimization design problem to test the ability of the metaheuristic methods by solving the minimization process of a weight tension/compression spring. The problem involves 3-dimensional search space; Wire diameter $W(x_1)$, mean coil diameter $d(x_2)$ and the number of active coils $L(x_3)$. It also presents three non-linear inequality constraints. The experimental

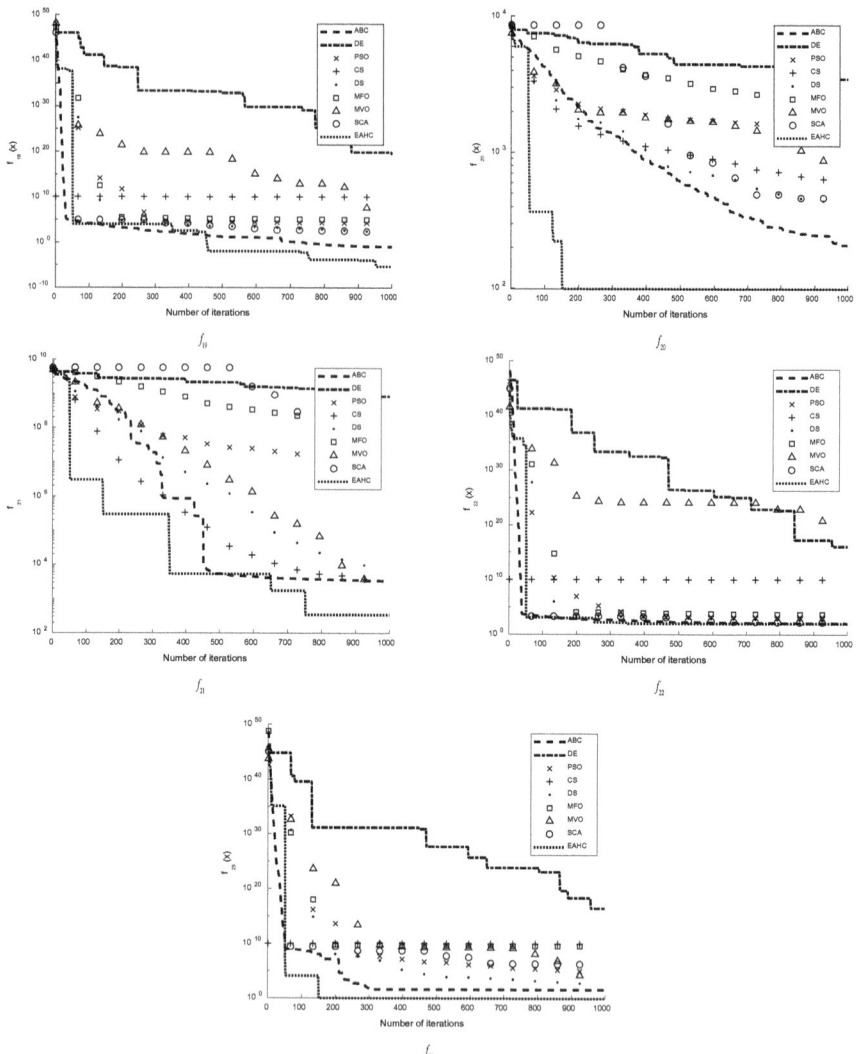

Fig. 8.4 (continued)

results for this experiment are presented in Table 8.8. Figure 8.6 presents a schematic of the tension/compression spring problem.

From Table 8.8, it can be shown that EA-HC produces worse fitness value than CS, DS, MFO, MVO and SCA. Even if these algorithms outperform the proposed method, the EA-HC method produces competitive results compared with the state-of-art algorithms. This effect can be produced by the efficient performance achieved by the dynamic balance among the coward and heroical evolutionary operators of EA-HC. Table 8.9 reports the statistical results for this experiment.

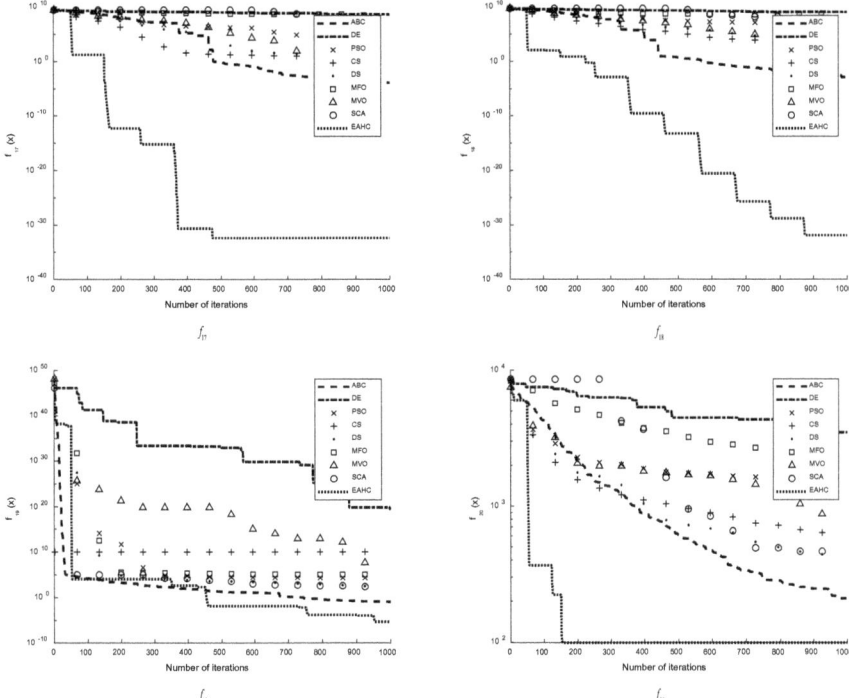

Fig. 8.4 (continued)

8.5.4.3 Welded Beam Design Problem

The welded beam optimization design problem corresponds to a complex engineering optimization task. The main objective of this approach consists of finding the lowest cost of a welded beam. The graphical description of the design problem is given in Fig. 8.7. The welded beam design process involves a 4-dimensional search space: Width $h(x_1)$, length $l(x_2)$, depth $t(x_3)$ and thickness $b(x_4)$. Additionally, this optimization task consists of 7 constraints described in Table 8.15 in Appendix 8.2. The numerical results are reported in Table 8.10.

From Table 8.10, it can be demonstrated the EA-HC is capable of achieving quite similar results than CS and MFO. To corroborate the experimental results from Tables 8.10 and 8.11 reports that EA-HC produces results with greater mean and standard deviation. However, this is not a limitation of the proposed approach, since it uses a changing framework overexploitation and exploration stages during the entire evolutionary process.

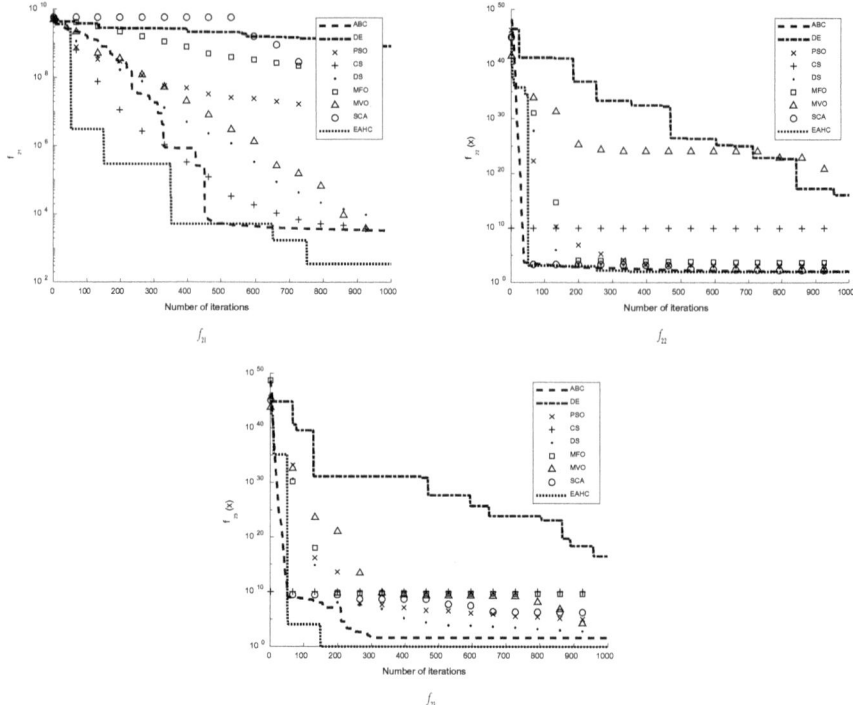

Fig. 8.4 (continued)

Fig. 8.5 Description of the three-bar truss design problem

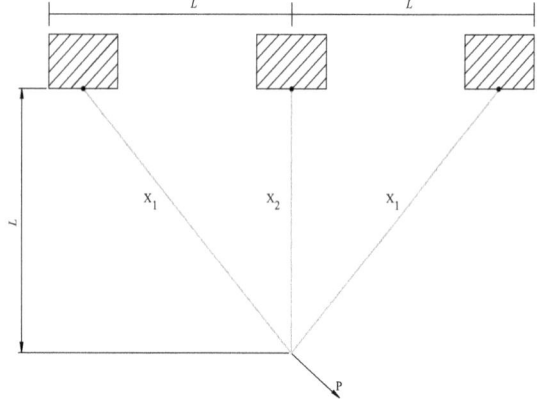

8.6   Conclusions

Agent-based modeling is conceived as a relatively new approach to model complex systems based on the interaction of agents using simple behaviors. As a consequence,

Table 8.6 Numerical results for the three-bar truss problem

Parameter	ABC	DE	PSO	CS	DS	MFO	MVO	SCA	EA-HC
x_1	0.8643	0.8708	0.8573	0.8698	0.8700	0.8702	0.8697	0.8730	0.8522
x_2	0.2278	0.2149	0.2710	0.2166	0.2162	0.2157	0.2167	0.2101	0.2528
$g_1(x)$	−1.49E−04	−3.60E−04	−0.0275	−2.224e−16	−6.274e−10	0	−2.561e−07	−5.740e−05	0.0001
$g_2(x)$	−1.6858	−1.7029	−1.6396	−1.7005	−1.7010	−1.7018	−1.7005	−1.7091	−1.6532
$g_3(x)$	−0.3143	−0.2975	−0.3878	−0.2995	−0.2990	−0.2982	−0.2995	−0.2909	0.3467
$f(x)$	279.7452	279.7472	280.5426	**279.7245**	**279.7245**	279.7246	**279.7245**	279.7312	**279.7245**

Table 8.7 Statistical results for the three-bar truss design problem

Algorithm	Worst	Mean	Std	Best
ABC	279.7849	279.7372	0.0125	279.7452
DE	280.1867	279.8096	0.0902	279.7472
PSO	282.8226	93.2079	29.9639	280.5426
CS	279.7245	279.7245	1.3843e−13	**279.7245**
DS	280.0971	279.7484	0.0677	**279.7245**
MFO	280.1945	279.7763	0.1028	279.7246
MVO	279.7270	279.7251	5.7728e−04	**279.7245**
SCA	282.8427	281.4059	1.5624	279.7312
EA-HC	279.9114	279.7491	0.0411	**279.7245**

the combination of simple interactions among agents will produce complex global behaviors. Since metaheuristic optimization techniques are conceived as stochastic search methods to obtain optimal solutions for complex optimization problems, the synergy among agent-based modeling and metaheuristic result in an interesting complementary search mechanism. Recently, many metaheuristic methods have been proposed based on a common set of rules regardless of the final global pattern obtained by individual interactions. In this chapter, an agent-based metaheuristic technique is presented. The proposed approach uses the model known as "Heroes and Cowards" to implement complex global search behaviors using simple agent interactions. The model incorporates a small set of rules to produce two emergent global patterns that can be considered in terms of the metaheuristic literature as exploration and exploitation stages.

In the complex system community, the model "Heroes and Cowards" is used to illustrate how simple rules can produce complex collective behaviors that are very difficult to reproduce by employing classical modeling techniques. The model produces complex global patterns of concentration and distribution through the interaction of agents that follows simple behavioral rules. The model is divided into two global behaviors; Coward (exploration) and heroical (exploitation). During the coward phase, agents are distributed along the space as a consequence of the scape process from the enemy. On the other hand, during the hero stage, agents semi-concentrate around positions marked by the agent distributions. Additionally, the model considers the use of a moderator to dynamically change among these two phases.

From a computational point of view, the proposed mechanism considers three elements; The initialization process, the coward and heroical evolutionary operators and a phase management mechanism to incorporate the dynamic change among exploration and exploitation stages.

The performance of the proposed method is numerically compared against several state-of-art metaheuristic methodologies evaluating 23 benchmark functions with

Table 8.8 Numerical results for the spring design problem

Parameter	ABC	DE	PSO	CS	DS	MFO	MVO	SCA	EA-HC
w	0.0500	0.0500	0.0516	0.0500	0.0500	0.0500	0.0500	0.0500	0.0518
d	0.4532	0.4278	0.3560	0.4800	0.4800	0.4800	0.4800	0.4796	0.4156
L	4.8539	5.8778	11.3295	4.0563	4.0577	4.0563	4.0580	4.0806	9.4088
$g_1(x)$	−0.0070	−0.0257	−0.000006	−1.547e−13	−4.834e−06	−0.0025	−6.662e−05	−0.0035	−0.3079
$g_2(x)$	−0.1001	−0.1896	−0.000013	−2.071e−14	−2.154e−04	−1.663e−15	−2.187e−04	−0.0015	−0.2361
$g_3(x)$	−6.0440	−5.5282	−4.0523	−6.5135	−6.5126	−6.4950	−6.5121	−6.4810	−3.4755
$g_4(x)$	−0.6645	−0.6815	−0.7282	−0.6467	−0.6467	−0.6467	−0.6467	−0.6469	−0.6884
$f(x)$	0.0078	0.0084	0.0127	**0.0073**	**0.0073**	**0.0073**	**0.0073**	**0.0073**	0.0076

Fig. 8.6 Description of the tension/compression spring design problem

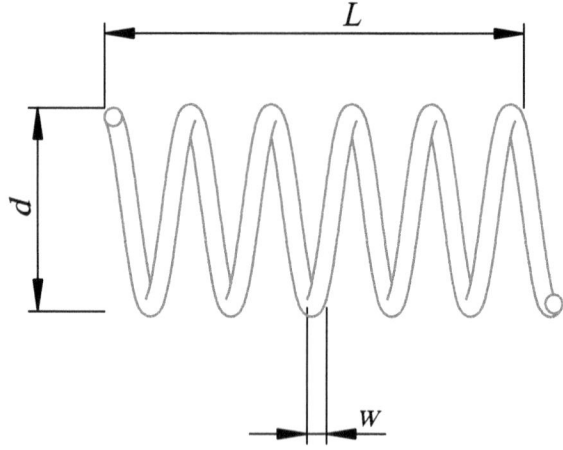

Table 8.9 Statistical results for the tension/compression spring design problem

Algorithm	Worst	Mean	Std	Best
ABC	0.0084	0.0076	0.0002	0.0078
DE	1.00E+06	2.00E+05	4.07E+05	0.0084
PSO	1.00E+06	1.67E+05	3.79E+05	0.0127
CS	0.0073	0.0073	2.3681e-12	**0.0073**
DS	0.0081	0.0074	2.2511e-04	**0.0073**
MFO	0.0073	0.0073	1.5149e-07	**0.0073**
MVO	1.00E+06	4.0000e+05	4.9827e+05	**0.0073**
SCA	0.0077	0.0074	9.4466e-05	**0.0073**
EA-HC	0.0237	0.0129	0.0031	0.0076

Fig. 8.7 Description of the Welded beam design problem

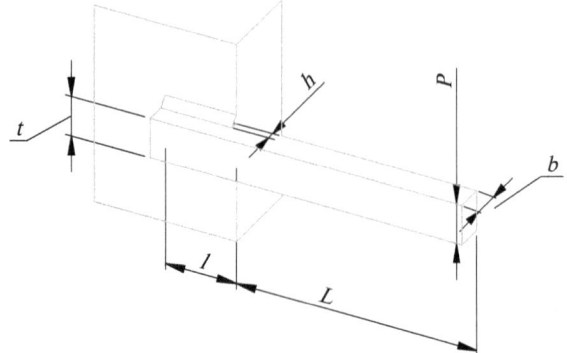

Table 8.10 Numerical results for the welded beam design problem

Parameter	ABC	DE	PSO	CS	DS	MFO	MVO	SCA	EA-HC
h	0.1993	0.1626	0.8120	0.2057	0.2057	0.2057	0.2045	0.2004	0.2057
l	3.8115	6.8239	6.4319	3.4705	3.4706	3.4705	3.5040	3.7893	3.4705
t	8.6442	8.7650	9.2837	9.0366	9.0385	9.0366	9.0375	9.0748	9.0366
b	0.2259	0.2346	0.7008	0.2057	0.2057	0.2057	0.2058	0.2062	0.2057
$g_1(x)$	−81.6604	−2977.1639	−1.16E+04	−0.0062	−0.0322	−0.0062	−23.2984	−616.4397	−0.0062
$g_2(x)$	−135.7300	−2035.9101	−2.17E+04	−0.1334	−14.1393	−0.1334	−9.1686	−315.5598	−0.1334
$g_3(x)$	−0.0265	−0.0720	0.1112	−1.283e−06	−5.025e−05	−1.283e−06	−0.0012	−0.0058	−1.286e−06
$g_4(x)$	−3.3229	−2.9372	1.4640	−3.4330	−3.4326	−3.4406	−3.4297	−3.3945	−3.4306
$g_5(x)$	−0.0743	−0.0376	−0.6870	−0.0807	−0.0807	−0.0807	−0.0795	−0.0754	−0.0807
$g_6(x)$	−0.2350	−0.2361	−0.2461	−0.2355	−0.2356	−0.2355	−0.2355	−0.2358	−0.2355
$g_7(x)$	−1706.7356	−2718.3738	−2.35E+05	−0.0157	−1.9242	−0.0157	−2.5351	−55.4759	−0.0157
$f(x)$	1.8403	2.2593	2.0118	**1.7249**	1.7252	**1.7249**	1.7278	1.7694	**1.7249**

Table 8.11 Statistical results for the welded beam design problem

Algorithm	Worst	Mean	Std	Best
ABC	2.0848	1.9269	0.0849	1.8403
DE	3.6161	2.6481	0.3445	2.2593
PSO	3.3931	2.1544	0.3594	2.0118
CS	1.7250	1.7249	2.2364e-05	**1.7249**
DS	2.2552	1.8165	0.1205	1.7252
MFO	2.3171	1.7854	0.1282	**1.7249**
MVO	1.8169	1.7412	0.0182	1.7278
SCA	1.9711	1.8519	0.0493	1.7694
EA-HC	2.3171	1.9850	0.2046	**1.7249**

different complexities. The experimental results indicate that the agent-based meta-heuristic approach overcomes its competitors in terms of accuracy and scalability. The remarkable performance of the proposed method is based on the dynamic change between exploration and exploitation during the entire optimization process.

Appendix 8.1: List of Benchmark Functions

See Table 8.12.

Appendix 8.2: Engineering Design Problems

See Tables 8.13, 8.14 and 8.15.

Table 8.12 Benchmark functions used in the experimental study

Function	S	Dimension	Minimum
$f_1(\mathbf{x}) = \sum_{i=1}^{n} x_i^2$	$[-100, 100]^n$	$n = 30$ $n = 100$	$\mathbf{x}^* = (0, \ldots, 0); f(\mathbf{x}^*) = 0$
$f_2(\mathbf{x}) = \sum_{i=1}^{n} |x_i| + \prod_{i=1}^{n} |x_i|$	$[-100, 100]^n$	$n = 30$ $n = 100$	$\mathbf{x}^* = (0, \ldots, 0); f(\mathbf{x}^*) = 0$
$f_3(\mathbf{x}) = \sum_{i=1}^{n} \left(\sum_{j=1}^{i} x_i^2 \right)^2$	$[-100, 100]^n$	$n = 30$ $n = 100$	$\mathbf{x}^* = (0, \ldots, 0); f(\mathbf{x}^*) = 0$
$f_4(\mathbf{x}) = \max\{|x_i|, 1 \leq i \leq n\}$	$[-100, 100]^n$	$n = 30$ $n = 100$	$\mathbf{x}^* = (0, \ldots, 0); f(\mathbf{x}^*) = 0$
$f_5(\mathbf{x}) = \sum_{i=1}^{n} \left[100(x_{i+1} - x_i^2)^2 + (x_i - 1)^2 \right]$	$[-30, 30]^n$	$n = 30$ $n = 100$	$\mathbf{x}^* = (1, \ldots, 1); f(\mathbf{x}^*) = n$
$f_6(\mathbf{x}) = \sum_{i=1}^{n} (\lfloor x_i + 0.5 \rfloor)^2$	$[-100, 100]^n$	$n = 30$ $n = 100$	$\mathbf{x}^* = (0, \ldots, 0); f(\mathbf{x}^*) = 0$
$f_7(\mathbf{x}) = \sum_{i=1}^{n} i x_i^4 + random(0, 1)$	$[-1.28, 1.28]^n$	$n = 30$ $n = 100$	$\mathbf{x}^* = (0, \ldots, 0); f(\mathbf{x}^*) = 0$
$f_8(\mathbf{x}) = \sum_{i=1}^{n} x_i + \sum_{i=1}^{n} \frac{x_i}{\left(-x_i + \sum_{i1=1}^{n} x_{i1} \right)^2}$	$[10e^{-6}, 2]^n$	$n = 30$ $n = 100$	$\mathbf{x}^* = \left(\frac{1}{n-1}, \ldots, \frac{1}{n-1} \right); f(\mathbf{x}^*) = \frac{2n}{n-1}$

(continued)

Table 8.12 (continued)

Function	S	Dimension	Minimum
$f_9(x) = \sum_{i=1}^{n} x_i^2 + \sum_{i=1}^{n} \dfrac{x_i^2}{(-x_i + \sum_{i1=1}^{n} x_{i1})^4}$	$[10e^{-6}, 2]^n$	$n = 30$ $n = 100$	$x^* = \left(\dfrac{1}{n-1}, \ldots, \dfrac{1}{n-1}\right); f(x^*) = \dfrac{2n}{(n-1)^2}$
$f_{10}(x) = \sum_{i=1}^{n} [x_i^2 - 10\cos(2\pi x_i) + 10]$	$[-5.12, 5.12]^n$	$n = 30$ $n = 100$	$x^* = (0, \ldots, 0); f(x^*) = 0$
$f_{11}(x) = -20\exp\left(-0.2\sqrt{\dfrac{1}{n}\sum_{i=1}^n x_i^2}\right) - \exp\left(\dfrac{1}{n}\sum_{i=1}^n \cos(2\pi x_i)\right) + 20 + e$	$[-32, 32]^n$	$n = 30$ $n = 100$	$x^* = (0, \ldots, 0); f(x^*) = 0$
$f_{12}(x) = -0.1\sum_{i=1}^n \cos(5\pi x_i) - \sum_{i=1}^n x_i^2$	$[-1, 1]^n$	$n = 30$ $n = 100$	$x^* = (0, \ldots, 0); f(x^*) =$ $-0.1 \times n$
$f_{13}(x) = -0.1\sum_{i=1}^n \cos(5\pi x_i) - \sum_{i=1}^n x_i^2$	$[-1, 1]^n$	$n = 30$ $n = 100$	$x^* = (0, \ldots, 0); f(x^*) =$ $-0.1 \times n$
$f_{14}(x) = \sum_{i=1}^n [x_i^2(2x_i^2 + x_{i+1} + 2) - x_i x_{i-1}(3x_i + 3x_{i-1} - x_{i+1} - 2)]^2 + \dfrac{1}{n^2}\sum_{i=1}^n (x_i - 1)^2 +$ $e; x_{n+1} = x_1, x_0 = x_n$	$[-1, 2]^n$	$n = 30$ $n = 100$	$x^* = (1, \ldots, 1); f(x^*) = 1$
$f_{15}(x) = -(n+1)e^{-10\sqrt{n}\left[\sum_{i=1}^n (x_i-1)^2\right]^{1/2}} + \sum_{i=1}^n x_i^2$	$[-2, 2]^n$	$n = 30$ $n = 100$	$x^* = (1, \ldots, 1); f(x^*) =$ -1

(continued)

Table 8.12 (continued)

Function	S	Dimension	Minimum
$f_{16}(x) = \sum_{i=1}^{n} -x_i \sin\left(\sqrt{|x_i|}\right)$	$[-500, 500]^n$	$n = 30$ $n = 100$	$x^* = (420, ..., 420); f(x^*) = -418.9829 \times n$
$f_{17}(x) = \frac{\pi}{n}\left\{10\sin(\pi y_i - 1) + \sum_{i=1}^{n-1}(y_i - 1)^2\left[1 + 10\sin^2(\pi y_{i+1}) + (y_n - 1)^2\right]\right\}$ $+ \sum_{i=1}^{n} u(x_i, 10, 100, 4);$ $y_i = 1 + \frac{x_i+1}{4};$ $u(x_i, a, k, m) = \begin{cases} k(x_i - a)^m, & x_i > a \\ 0, & -a < x_i < a \\ k(-x_i - a)^m, & x_i < -a \end{cases}$	$[-50, 50]^n$	$n = 30$ $n = 100$	$x^* = (0, ..., 0); f(x^*) = 0$
$f_{18}(x) =$ $0.1\left\{\sin^2(3\pi x_i) + \sum_{i=1}^{n}(x_i - 1)^2\left[1 + \sin^2(3\pi x_1 + 1)\right] + (x_n - 1)^2\left[1 + \sin^2(2\pi x_n)\right]\right\} +$ $u(x_i, 5, 100, 4)$	$[-50, 50]^n$	$n = 30$ $n = 100$	$x^* = (1, ..., 1); f(x^*) = 0$
$f_{19}(x) = f_1(x) + f_2(x) + f_{10}(x)$	$[-100, 100]^n$	$n = 30$ $n = 100$	$x^* = (0, ..., 0); f(x^*) = 0$
$f_{20}(x) = f_5(x) + f_{10}(x) + f_{12}(x)$	$[-100, 100]^n$	$n = 30$ $n = 100$	$x^* = (0, ..., 0); f(x^*) =$ $n - 1$

(continued)

Table 8.12 (continued)

Function	S	Dimension	Minimum
$f_{21}(x) = f_3(x) + f_5(x) + f_{11}(x) + f_{18}(x)$	$[-100, 100]^n$	$n = 30$ $n = 100$	$x^* = (0, ..., 0); f(x^*) = (1.1 \times n) - 1$
$f_{22}(x) = f_2(x) + f_5(x) + f_{10}(x) + f_{11}(x) + f_{12}(x)$	$[-100, 100]^n$	$n = 30$ $n = 100$	$x^* = (0, ..., 0); f(x^*) = n - 1$
$f_{23}(x) = f_1(x) + f_2(x) + f_{11}(x) + f_{16}(x) + f_{17}(x)$	$[-100, 100]^n$	$n = 30$ $n = 100$	$x^* = (1, ..., 1); f(x^*) = -3 \times n$

Table 8.13 Three-bar truss design description used in the experimental study

Function	S	Dimension	Constraints
$f(x) = \left(2\sqrt{2}x_1 + x_2\right) \times$ $LL = 100$ cm, $P =$ $2\,\text{kN/cm}^2$, $\sigma = 2\,\text{kN/cm}^2$	$0 \le x_i \le 1, i = 1, 2$	$n = 2$	$g_1(x) = \frac{\sqrt{2}x_1 + x_2}{\sqrt{2}x_1^2 + 2x_1x_2} P - \sigma \le$ $0 g_2(x) = \frac{x_2}{\sqrt{2}x_1^2 + 2x_1x_2} P - \sigma \le$ $0 g_3(x) = \frac{1}{\sqrt{2}x_2 + x_1} P - \sigma \le 0$

Table 8.14 Tension/compression spring design description used in the experimental study

Function	S	Dimension	Constraints
$f(x) = (x_3 + 2)x_2x_1^2$	$0.05 \le x_1 \le 20.25 \le x_2 \le 1.32 \le x_3 \le 15$	$n = 3$	0

Table 8.15 Welded beam design description used in the experimental study

Function	S	Dimension	Constraints
$f(x) =$ $1.10471x_1^2x_2 +$ $0.04811x_3x_4(14 + x_2)$	$0.1 \le x_i \le 2,$ $i = 1, 40.1 \le$ $x_i \le 10,$ $i = 2, 3$	$n = 4$	$\tau(x) =$ $\sqrt{(\tau')^2 + 2\tau\tau'' \frac{x_2}{2R} + (\tau'')^2}, \tau' =$ $\frac{P}{\sqrt{2}x_1x_2}, \tau'' = \frac{MR}{J}, M =$ $P\left(L + \frac{x_2}{2}\right), R =$ $\sqrt{\frac{x_2^2}{2} + \left(\frac{x_1 + x_3}{2}\right)^2}, \delta(x) = \frac{4PL^3}{Ex_3^3x_4}, J =$ $2\left[\sqrt{2}x_1x_2\left\{\frac{x_2^2}{12} + \left(\frac{x_1+x_3}{2}\right)^2\right\}\right], \sigma(x) =$ $\frac{6PL}{x_4x_3^2}, P_c(x) =$ $\frac{4.013E\sqrt{\frac{x_3^2x_4^6}{36}}}{L^2}\left(1 - \frac{x_3}{2L}\sqrt{\frac{E}{4G}}\right)$

References

1. Wilensky U, Rand W (2010) An introduction to agent-based modeling modeling natural, social, and engineered complex systems with NetLogo. MIT Press
2. Banisch S (2016) Markov chain aggregation for agent-based models. Springer
3. Macal1 C, North M (2010) Tutorial on agent-based modelling and simulation. J Simul 4:151–162
4. Bagni R, Berchi R, Cariello P (2002) A comparison of simulation models applied to epidemics. J Artif Soc Social Simul 5(3)
5. Macal CM (2004) Emergent structures from trust relationships in supply chains. In: Macal C, Sallach D, North M (eds) Proceedings of agent 2004: conference on social dynamics: interaction, reflexivity and emergence. Argonne National Laboratory, Chicago, IL, 7–9 Oct, pp 743–760
6. Arthur WB, Durlauf SN, Lane DA (eds) (1997) The economy as an evolving complex system II. SFI studies in the sciences of complexity. Addison-Wesley, Reading, MA
7. Folcik VA, An GC, Orosz CG (2007) The basic immune simulator: an agent-based model to study the interactions between innate and adaptive immunity. Theoret Biol Med Model 4(39). http://www.tbiomed.com/content/4/1/39
8. Kohler TA, Gumerman GJ, Reynolds RG (2005) Simulating ancient societies. Sci Am 293(1):77–84
9. North M et al (2009) Multi-scale agent-based consumer market modeling. Complexity 15:37–47
10. Bonabeau E (2012). http://www.icosystem.com/labsdemos/the-game/
11. Sweeney LB, Meadows D (2010) The systems thinking playbook: exercises to stretch and build learning and systems thinking capabilities. Chelsea Green Publishing, White River Junction, VT
12. Bonabeau E, Meyer C (2001) Swarm intelligence: a whole new way to think about business. Harv Bus Rev 5:107–114
13. Sörensen K (2015) Metaheuristics—the metaphor exposed. Int Trans Oper Res 22(1):3–18
14. Fausto F, Reyna-Orta A, Cuevas E, Andrade ÁG, Perez-Cisneros M (2019) From ants to whales: metaheuristics for all tastes. Springer Netherlands (2019). https://doi.org/10.1007/s10462-018-09676-2
15. Hussain K, Mohd Salleh MN, Cheng S, Shi Y (2018) Metaheuristic research: a comprehensive survey. Artif Intell Rev 1–43. https://doi.org/10.1007/s10462-017-9605-z
16. Banharnsakun A, Achalakul T, Sirinaovakul B (2011) The best-so-far selection in Artificial Bee Colony algorithm. Appl Soft Comput 11:2888–2901
17. Kennedy J, Eberhart R (1995) Particle swarm optimization. Neural networks. Proceedings. IEEE international conference, vol 4, 1942–1948
18. Poli R, Kennedy J, Blackwell T (2007) Particle swarm optimization. Swarm Intell 1:33–57
19. Marini F, Walczak B (2015) Particle swarm optimization (PSO). A tutorial. Chemom Intell Lab Syst 149:153–165
20. Cuevas E, Cienfuegos M, Zaldívar D, Pérez-cisneros M (2013) A swarm optimization algorithm inspired in the behavior of the social-spider. Expert Syst Appl 40:6374–6384
21. Askarzadeh A (2016) A novel metaheuristic method for solving constrained engineering optimization problems: crow search algorithm. Comput Struct 169:1–12. https://doi.org/10.1016/j.compstruc.2016.03.001
22. Díaz P, Pérez M, Cuevas E, Avalos O, Gálvez J, Hinojosa S, Zaldivar D (2018) An improved crow search algorithm applied to energy problems 1–23. https://doi.org/10.3390/en11030571
23. Mirjalili S, Mirjalili SM, Lewis A (2014) Grey wolf optimizer. Adv Eng Softw 69:46–61
24. Yang X-S (2010) A new metaheuristic Bat-inspired algorithm. Stud Comput Intell 284:65–74
25. Yang XS, Deb S (2009) Cuckoo search via Lévy flights. In: 2009 world congress on nature & biologically inspired computing. NABIC 2009 proceedings, pp 210–214

26. Yang X-S (2009) Firefly algorithms for multimodal optimization. In: Lecture Notes in Computer Science. Lecture notes in artificial intelligence and lecture notes in bioinformatics, pp 169–178. https://doi.org/10.1007/978-3-642-04944-6_14
27. Yang X-S (2010) Firefly algorithm, Lévy flights and global optimization. In: Research and development in intelligent systems, vol XXVI. Springer London, London, pp 209–218. https://doi.org/10.1007/978-1-84882-983-1_15
28. Yang XS (2009) Firefly algorithms for multimodal optimization. In: Lecture Notes in Computer Science. Lecture notes in artificial intelligence and lecture notes in bioinformatics. 5792 LNCS, pp 169–178
29. Rashedi E, Nezamabadi-pour H, Saryazdi S (2009) GSA: a gravitational search algorithm. Inf Sci (NY) 179:2232–2248
30. Kirkpatrick S, Gelatt CD, Vecchi MP (1983) Optimization by simulated annealing. Science (80-), 220, 671–680. https://doi.org/10.1126/science.220.4598.671
31. Rutenbar RA (1989) Simulated annealing algorithms: an overview. IEEE Circuits Dev Mag 5:19–26
32. Siddique N, Adeli H (2016) Simulated annealing, its variants and engineering applications. Int J Artif Intell Tools 25:1630001
33. Birbil ŞI, Fang SC (2003) An electromagnetism-like mechanism for global optimization. J Glob Optim 25:263–282
34. Karaboga D, Basturk B (2008) On the performance of artificial bee colony (ABC) algorithm. Appl Soft Comput J 8:687–697
35. Cuevas E, Echavarría A, Ramírez-Ortegón MA (2013) An optimization algorithm inspired by the states of matter that improves the balance between exploration and exploitation. Appl Intell 40:256–272. https://doi.org/10.1007/s10489-013-0458-0
36. Valdivia-Gonzalez A, Zaldívar D, Fausto F, Camarena O, Cuevas E, Perez-Cisneros M (2017) A states of matter search-based approach for solving the problem of intelligent power allocation in plug-in hybrid electric vehicles. Energies 10:92. https://doi.org/10.3390/en10010092
37. Eskandar H, Sadollah A, Bahreininejad A, Hamdi M (2012) Water cycle algorithm—a novel metaheuristic optimization method for solving constrained engineering optimization problems. Comput Struct 110–111:151–166
38. Tang KS, Man KF, Kwong S, He Q (1996) Genetic algorithms and their applications. IEEE Signal Process Mag 13:22–37. https://doi.org/10.1109/79.543973
39. Beyer H-G, Beyer H-G, Schwefel H-P, Schwefel H-P (2002) Evolution strategies—a comprehensive introduction. Nat Comput 1:3–52. https://doi.org/10.1023/A:1015059928466
40. Bäck T, Hoffmeister F, Schwefel H-P (1991) A survey of evolution strategies. In: Proceedings of the fourth international conference on genetic algorithms, p 8. 10.1.1.42.3375
41. Hansen N (2016) The CMA evolution strategy: a tutorial. 102:75–102
42. Zhang J, Sanderson AC (2007) JADE: self-adaptive differential evolution with fast and reliable convergence performance. In: 2007 IEEE congress on evolutionary computation, CEC 2007
43. Storn R, Price K (1997) Differential evolution—a simple and efficient heuristic for global optimization over continuous spaces. J Glob Optim 11:341–359
44. Chan V, Son Y, Macal C (2010) Agent-based simulation tutorial—simulation of emergent behavior and differences between agent-based simulation and discrete-event simulation. In: Johansson B, Jain S, Montoya-Torres J, Hugan J, Yücesan E (eds) Proceedings of the 2010 winter simulation conference
45. Schelling TC (1971) Dynamic models of segregation. J Math Sociol 1(2):143–186
46. Sayama H (2015) Introduction to the modeling and analysis of complex systems. Open SUNY textbooks, Milne Library
47. Anescu (2017) Further scalable test functions for multidimensional continuous optimization
48. Li MD, Zhao H, Weng XW, Han T (2016) A novel nature-inspired algorithm for optimization: virus colony search. Adv Eng Softw 92:65–88
49. Yang X-S (2010) Wiley InterScience (Online service). Engineering optimization: an introduction with metaheuristic applications. Wiley

50. Karaboga D (2005) An idea based on honey bee swarm for numerical optimization. TechnicalReport-TR06. Engineering Faculty, Computer Engineering Department, Erciyes University
51. Civicioglu P (2012) Transforming geocentric cartesian coordinates to geodetic coordinates by using differential search algorithm. Comput Geosci 46:229–247
52. Mirjalili S (2015) Moth-flame optimization algorithm: a novel nature-inspired heuristic paradigm. Knowl Based Syst 89:228–249
53. Mirjalili S, Mirjalili SM, Hatamlou A (2016) Multi-verse optimizer: a nature-inspired algorithm for global optimization. Neural Comput Appl 27(2):495–513
54. Mirjalili S (2016) SCA: a Sine Cosine algorithm for solving optimization problems. Knowl Based Syst 96:120–133
55. Rashedi E, Nezamabadi-pour H, Saryazdi S (2009) GSA: a gravitational search algorithm. Inf Sci (NY) 179(13):2232–2248
56. Yu S, Zhu S, Ma Y, Mao D (2015) A variable step size firefly algorithm for numerical optimization. Appl Math Comput 263:214–220
57. Yang X-S, Karamanoglu M, He X (2014) Flower pollination algorithm: a novel approach for multiobjective optimization. Eng Optim 46(9):1222–1237
58. Wilcoxon F (1945) Individual comparisons by ranking methods. Biometrics 80–83